LONDON MATHEMATICAL SOCIETY LECTURE NOTE SERIES

Managing Editor: Professor M. Reid, Mathematics Institute,
University of Warwick, Coventry CV4 7AL, United Kingdom

The titles below are available from booksellers, or from Cambridge University Press at
http://www.cambridge.org/mathematics

311 Groups: topological, combinatorial and arithmetic aspects, T.W. MÜLLER (ed)
312 Foundations of computational mathematics, Minneapolis 2002, F. CUCKER *et al* (eds)
313 Transcendental aspects of algebraic cycles, S. MÜLLER-STACH & C. PETERS (eds)
314 Spectral generalizations of line graphs, D. CVETKOVIĆ, P. ROWLINSON & S. SIMIĆ
315 Structured ring spectra, A. BAKER & B. RICHTER (eds)
316 Linear logic in computer science, T. EHRHARD, P. RUET, J.-Y. GIRARD & P. SCOTT (eds)
317 Advances in elliptic curve cryptography, I.F. BLAKE, G. SEROUSSI & N.P. SMART (eds)
318 Perturbation of the boundary in boundary-value problems of partial differential equations, D.HENRY
319 Double affine Hecke algebras, I. CHEREDNIK
320 L-functions and Galois representations, D. BURNS, K. BUZZARD & J. NEKOVÁŘ (eds)
321 Surveys in modern mathematics, V. PRASOLOV & Y. ILYASHENKO (eds)
322 Recent perspectives in random matrix theory and number theory, F. MEZZADRI & N.C. SNAITH (eds)
323 Poisson geometry, deformation quantisation and group representations, S. GUTT *et al* (eds)
324 Singularities and computer algebra, C. LOSSEN & G. PFISTER (eds)
325 Lectures on the Ricci flow, P. TOPPING
326 Modular representations of finite groups of Lie type, J.E. HUMPHREYS
327 Surveys in combinatorics 2005, B.S. WEBB (ed)
328 Fundamentals of hyperbolic manifolds, R. CANARY, D. EPSTEIN & A. MARDEN (eds)
329 Spaces of Kleinian groups, Y. MINSKY, M. SAKUMA & C. SERIES (eds)
330 Noncommutative localization in algebra and topology, A. RANICKI (ed)
331 Foundations of computational mathematics, Santander 2005, L.M PARDO, A. PINKUS, E. SÜLI & M.J. TODD (eds)
332 Handbook of tilting theory, L. ANGELERI HÜGEL, D. HAPPEL & H. KRAUSE (eds)
333 Synthetic differential geometry (2nd Edition), A. KOCK
334 The Navier–Stokes equations, N. RILEY & P. DRAZIN
335 Lectures on the combinatorics of free probability, A. NICA & R. SPEICHER
336 Integral closure of ideals, rings, and modules, I. SWANSON & C. HUNEKE
337 Methods in Banach space theory, J.M.F. CASTILLO & W.B. JOHNSON (eds)
338 Surveys in geometry and number theory, N. YOUNG (ed)
339 Groups St Andrews 2005 I, C.M. CAMPBELL, M.R. QUICK, E.F. ROBERTSON & G.C. SMITH (eds)
340 Groups St Andrews 2005 II, C.M. CAMPBELL, M.R. QUICK, E.F. ROBERTSON & G.C. SMITH (eds)
341 Ranks of elliptic curves and random matrix theory, J.B. CONREY, D.W. FARMER, F. MEZZADRI & N.C. SNAITH (eds)
342 Elliptic cohomology, H.R. MILLER & D.C. RAVENEL (eds)
343 Algebraic cycles and motives I, J. NAGEL & C. PETERS (eds)
344 Algebraic cycles and motives II, J. NAGEL & C. PETERS (eds)
345 Algebraic and analytic geometry, A. NEEMAN
346 Surveys in combinatorics 2007, A. HILTON & J. TALBOT (eds)
347 Surveys in contemporary mathematics, N. YOUNG & Y. CHOI (eds)
348 Transcendental dynamics and complex analysis, P.J. RIPPON & G.M. STALLARD (eds)
349 Model theory with applications to algebra and analysis I, Z. CHATZIDAKIS, D. MACPHERSON, A. PILLAY & A. WILKIE (eds)
350 Model theory with applications to algebra and analysis II, Z. CHATZIDAKIS, D. MACPHERSON, A. PILLAY & A. WILKIE (eds)
351 Finite von Neumann algebras and masas, A.M. SINCLAIR & R.R. SMITH
352 Number theory and polynomials, J. MCKEE & C. SMYTH (eds)
353 Trends in stochastic analysis, J. BLATH, P. MÖRTERS & M. SCHEUTZOW (eds)
354 Groups and analysis, K. TENT (ed)
355 Non-equilibrium statistical mechanics and turbulence, J. CARDY, G. FALKOVICH & K. GAWEDZKI
356 Elliptic curves and big Galois representations, D. DELBOURGO
357 Algebraic theory of differential equations, M.A.H. MACCALLUM & A.V. MIKHAILOV (eds)
358 Geometric and cohomological methods in group theory, M.R. BRIDSON, P.H. KROPHOLLER & I.J. LEARY (eds)
359 Moduli spaces and vector bundles, L. BRAMBILA-PAZ, S.B. BRADLOW, O. GARCÍA-PRADA & S. RAMANAN (eds)
360 Zariski geometries, B. ZILBER
361 Words: Notes on verbal width in groups, D. SEGAL
362 Differential tensor algebras and their module categories, R. BAUTISTA, L. SALMERÓN & R. ZUAZUA
363 Foundations of computational mathematics, Hong Kong 2008, F. CUCKER, A. PINKUS & M.J. TODD (eds)
364 Partial differential equations and fluid mechanics, J.C. ROBINSON & J.L. RODRIGO (eds)
365 Surveys in combinatorics 2009, S. HUCZYNSKA, J.D. MITCHELL & C.M. RONEY-DOUGAL (eds)
366 Highly oscillatory problems, B. ENGQUIST, A. FOKAS, E. HAIRER & A. ISERLES (eds)
367 Random matrices: High dimensional phenomena, G. BLOWER
368 Geometry of Riemann surfaces, F.P. GARDINER, G. GONZÁLEZ-DIEZ & C. KOUROUNIOTIS (eds)
369 Epidemics and rumours in complex networks, M. DRAIEF & L. MASSOULIÉ
370 Theory of p-adic distributions, S. ALBEVERIO, A.YU. KHRENNIKOV & V.M. SHELKOVICH
371 Conformal fractals, F. PRZYTYCKI & M. URBAŃSKI

372 Moonshine: The first quarter century and beyond, J. LEPOWSKY, J. MCKAY & M.P. TUITE (eds)
373 Smoothness, regularity and complete intersection, J. MAJADAS & A. G. RODICIO
374 Geometric analysis of hyperbolic differential equations: An introduction, S. ALINHAC
375 Triangulated categories, T. HOLM, P. JØRGENSEN & R. ROUQUIER (eds)
376 Permutation patterns, S. LINTON, N. RUŠKUC & V. VATTER (eds)
377 An introduction to Galois cohomology and its applications, G. BERHUY
378 Probability and mathematical genetics, N. H. BINGHAM & C. M. GOLDIE (eds)
379 Finite and algorithmic model theory, J. ESPARZA, C. MICHAUX & C. STEINHORN (eds)
380 Real and complex singularities, M. MANOEL, M.C. ROMERO FUSTER & C.T.C WALL (eds)
381 Symmetries and integrability of difference equations, D. LEVI, P. OLVER, Z. THOMOVA & P. WINTERNITZ (eds)
382 Forcing with random variables and proof complexity, J. KRAJÍČEK
383 Motivic integration and its interactions with model theory and non-Archimedean geometry I, R. CLUCKERS, J. NICAISE & J. SEBAG (eds)
384 Motivic integration and its interactions with model theory and non-Archimedean geometry II, R. CLUCKERS, J. NICAISE & J. SEBAG (eds)
385 Entropy of hidden Markov processes and connections to dynamical systems, B. MARCUS, K. PETERSEN & T. WEISSMAN (eds)
386 Independence-friendly logic, A.L. MANN, G. SANDU & M. SEVENSTER
387 Groups St Andrews 2009 in Bath I, C.M. CAMPBELL *et al* (eds)
388 Groups St Andrews 2009 in Bath II, C.M. CAMPBELL *et al* (eds)
389 Random fields on the sphere, D. MARINUCCI & G. PECCATI
390 Localization in periodic potentials, D.E. PELINOVSKY
391 Fusion systems in algebra and topology, M. ASCHBACHER, R. KESSAR & B. OLIVER
392 Surveys in combinatorics 2011, R. CHAPMAN (ed)
393 Non-abelian fundamental groups and Iwasawa theory, J. COATES *et al* (eds)
394 Variational problems in differential geometry, R. BIELAWSKI, K. HOUSTON & M. SPEIGHT (eds)
395 How groups grow, A. MANN
396 Arithmetic differential operators over the p-adic integers, C.C. RALPH & S.R. SIMANCA
397 Hyperbolic geometry and applications in quantum chaos and cosmology, J. BOLTE & F. STEINER (eds)
398 Mathematical models in contact mechanics, M. SOFONEA & A. MATEI
399 Circuit double cover of graphs, C.-Q. ZHANG
400 Dense sphere packings: a blueprint for formal proofs, T. HALES
401 A double Hall algebra approach to affine quantum Schur–Weyl theory, B. DENG, J. DU & Q. FU
402 Mathematical aspects of fluid mechanics, J.C. ROBINSON, J.L. RODRIGO & W. SADOWSKI (eds)
403 Foundations of computational mathematics, Budapest 2011, F. CUCKER, T. KRICK, A. PINKUS & A. SZANTO (eds)
404 Operator methods for boundary value problems, S. HASSI, H.S.V. DE SNOO & F.H. SZAFRANIEC (eds)
405 Torsors, étale homotopy and applications to rational points, A.N. SKOROBOGATOV (ed)
406 Appalachian set theory, J. CUMMINGS & E. SCHIMMERLING (eds)
407 The maximal subgroups of the low-dimensional finite classical groups, J.N. BRAY, D.F. HOLT & C.M. RONEY-DOUGAL
408 Complexity science: The Warwick master's course, R. BALL, V. KOLOKOLTSOV & R.S. MACKAY (eds)
409 Surveys in combinatorics 2013, S.R. BLACKBURN, S. GERKE & M. WILDON (eds)
410 Representation theory and harmonic analysis of wreath products of finite groups, T. CECCHERINI-SILBERSTEIN, F. SCARABOTTI & F. TOLLI
411 Moduli spaces, L. BRAMBILA-PAZ, O. GARCÍA-PRADA, P. NEWSTEAD & R.P. THOMAS (eds)
412 Automorphisms and equivalence relations in topological dynamics, D.B. ELLIS & R. ELLIS
413 Optimal transportation, Y. OLLIVIER, H. PAJOT & C. VILLANI (eds)
414 Automorphic forms and Galois representations I, F. DIAMOND, P.L. KASSAEI & M. KIM (eds)
415 Automorphic forms and Galois representations II, F. DIAMOND, P.L. KASSAEI & M. KIM (eds)
416 Reversibility in dynamics and group theory, A.G. O'FARRELL & I. SHORT
417 Recent advances in algebraic geometry, C.D. HACON, M. MUSTAŢĂ & M. POPA (eds)
418 The Bloch–Kato conjecture for the Riemann zeta function, J. COATES, A. RAGHURAM, A. SAIKIA & R. SUJATHA (eds)
419 The Cauchy problem for non-Lipschitz semi-linear parabolic partial differential equations, J.C. MEYER & D.J. NEEDHAM
420 Arithmetic and geometry, L. DIEULEFAIT *et al* (eds)
421 O-minimality and Diophantine geometry, G.O. JONES & A.J. WILKIE (eds)
422 Groups St Andrews 2013, C.M. CAMPBELL *et al* (eds)
423 Inequalities for graph eigenvalues, Z. STANIĆ
424 Surveys in combinatorics 2015, A. CZUMAJ *et al* (eds)
425 Geometry, topology and dynamics in negative curvature, C.S. ARAVINDA, F.T. FARRELL & J.-F. LAFONT (eds)
426 Lectures on the theory of water waves, T. BRIDGES, M. GROVES & D. NICHOLLS (eds)
427 Recent advances in Hodge theory, M. KERR & G. PEARLSTEIN (eds)
428 Geometry in a Fréchet context, C. T. J. DODSON, G. GALANIS & E. VASSILIOU
429 Sheaves and functions modulo p, L. TAELMAN
430 Recent progress in the theory of the Euler and Navier–Stokes equations, J.C. ROBINSON, J.L. RODRIGO, W. SADOWSKI & A. VIDAL-LÓPEZ (eds)
431 Harmonic and subharmonic function theory on the hyperbolic ball, M. STOLL
432 Topics in graph automorphisms and reconstruction (2nd Edition), J. LAURI & R. SCAPELLATO
433 Regular and irregular holonomic D-modules, M. KASHIWARA & P. SCHAPIRA
434 Analytic semigroups and semilinear initial boundary value problems (2nd Edition), K. TAIRA
435 Graded rings and graded Grothendieck groups, R. HAZRAT

London Mathematical Society Lecture Note Series: 431

Harmonic and Subharmonic Function Theory on the Hyperbolic Ball

MANFRED STOLL
University of South Carolina

CAMBRIDGE
UNIVERSITY PRESS

CAMBRIDGE
UNIVERSITY PRESS

Shaftesbury Road, Cambridge CB2 8EA, United Kingdom

One Liberty Plaza, 20th Floor, New York, NY 10006, USA

477 Williamstown Road, Port Melbourne, VIC 3207, Australia

314–321, 3rd Floor, Plot 3, Splendor Forum, Jasola District Centre, New Delhi – 110025, India

103 Penang Road, #05–06/07, Visioncrest Commercial, Singapore 238467

Cambridge University Press is part of Cambridge University Press & Assessment, a department of the University of Cambridge.

We share the University's mission to contribute to society through the pursuit of education, learning and research at the highest international levels of excellence.

www.cambridge.org
Information on this title: www.cambridge.org/9781107541481

First published 2016

A catalogue record for this publication is available from the British Library

Library of Congress Cataloging-in-Publication data
Names: Stoll, Manfred.
Title: Harmonic and subharmonic function theory on the hyperbolic ball / Manfred Stoll, University of South Carolina.
Description: Cambridge : Cambridge University Press, 2016. | Series: London Mathematical Society lecture note series ; 431 | Includes bibliographical references and index.
Identifiers: LCCN 2015049530 | ISBN 9781107541481 (pbk.)
Subjects: LCSH: Harmonic functions. | Subharmonic functions. | Hyperbolic spaces.
Classification: LCC QA405 .S76 2016 | DDC 515/.53–dc23
LC record available at http://lccn.loc.gov/2015049530

ISBN 978-1-107-54148-1 Paperback

To Mary Lee

Contents

Preface

The intent of these notes is to provide a detailed and comprehensive treatment of harmonic and subharmonic function theory on hyperbolic space in $\mathbb{R}^n$. Although our primary emphasis will be in the setting of the unit ball $\mathbb{B}$ with hyperbolic metric ds given by

$$ds = \frac{2|dx|}{1 - |x|^2}, \tag{1}$$

we will also consider the analogue of many of the results in the hyperbolic half-space $\mathbb{H}$. Undoubtedly some of the results are known, either in the setting of rank one noncompact symmetric spaces (e.g. [38]), or more generally, in Riemannian spaces (e.g. [13]). An excellent introduction to harmonic function theory on noncompact symmetric spaces can be found in the survey article [47] by A. Koranyi. The 1973 paper by K. Minemura [57] provides an introduction to harmonic function theory on real hyperbolic space considered as a rank one noncompact symmetric space. Other contributions to the subject area in this setting will be indicated in the text.

With the goal of making these notes accessible to a broad audience, our approach does not require any knowledge of Lie groups and only a limited knowledge of differential geometry. The development of the theory is analogous to the approach taken by W. Rudin [72] and by the author [84] in their development of Möbius invariant harmonic function theory on the hermitian ball in $\mathbb{C}^n$. Although our primary emphasis is on harmonic function theory on the ball, we do include many relevant results for the hyperbolic upper half-space $\mathbb{H}$, both in the text and in the exercises. With only one or two exceptions, the notes are self-contained with the only prerequisites being a standard beginning graduate course in real analysis.

In Chapter 1 we provide a brief review of Möbius transformation in $\mathbb{R}^n$. This is followed in Chapter 2 by a characterization of the group $\mathcal{M}(\mathbb{B})$ of

Möbius self-maps of the unit ball $\mathbb{B}$ in $\mathbb{R}^n$. As in [72] we define a family $\{\varphi_a : a \in \mathbb{B}\}$ of Möbius transformations of $\mathbb{B}$ satisfying $\varphi_a(0) = a$, $\varphi_a(a) = 0$, and $\varphi_a(\varphi_a(x)) = x$ for all $x \in \mathbb{B}$. Furthermore, for every $\psi \in \mathcal{M}(\mathbb{B})$, it is proved that there exists $a \in \mathbb{B}$ and an orthogonal transformation A such that $\psi = A\varphi_a$. When $n = 2$, the mappings φ_a correspond to the usual analytic Möbius transformations of the unit disc $\mathbb{D}$ given by

$$\varphi_a(z) = \frac{a - z}{1 - \bar{a}z}. \tag{2}$$

Some of the properties of the mappings $\{\varphi_a\}$ and of functions in $\mathcal{M}(\mathbb{B})$ are developed in Section 2.1. In this chapter we also introduce the hyperbolic metric in $\mathbb{B}$ and in the hyperbolic half-space $\mathbb{H}$. Most of the results of these two sections are contained in the works of L. V. Ahlfors [4], [5] , and the text by A. F. Beardon [11].

In Chapter 3 we derive the Laplacian, gradient, and measure on $\mathbb{B}$ that are invariant under $\mathcal{M}(\mathbb{B})$. Even though the formula for the Laplacian can be derived from the hyperbolic metric, we will follow the approach of W. Rudin [72, Chapter 4]. For $f \in C^2(\mathbb{B})$ we define $\Delta_h f$ by

$$\Delta_h f(a) = \Delta(f \circ \varphi_a)(0),$$

where Δ is the usual Laplacian in $\mathbb{R}^n$. The operator Δ_h is shown to satisfy $\Delta_h(f \circ \psi)(x) = (\Delta_h f)(\psi(x))$ for all $\psi \in \mathcal{M}(\mathbb{B})$. Furthermore, an explicit computation gives

$$\Delta_h f(x) = (1 - |x|^2)^2 \Delta f(x) + 2(n-2)(1 - |x|^2)\langle x, \nabla f(x)\rangle,$$

where ∇f is the Euclidean gradient of the function f. In this chapter it is also proved that the Green's function for Δ_h is given by $G_h(x, y) = g(|\varphi_x(y)|)$, where g is the radial function on $\mathbb{B}$ defined by

$$g(r) = \frac{1}{n}\int_r^1 \frac{(1 - s^2)^{n-2}}{s^{n-1}}\, ds.$$

In Theorem 3.3.1 we prove that for $\psi \in \mathcal{M}(\mathbb{B})$, the Jacobian J_ψ of the mapping ψ satisfies

$$|J_\psi(x)| = \frac{(1 - |\psi(x)|^2)^n}{(1 - |x|^2)^n}.$$

From this it now follows that the Möbius invariant measure τ on $\mathbb{B}$ is given by

$$d\tau(x) = (1 - |x|^2)^{-n} d\nu(x),$$

where ν is the normalized volume measure on $\mathbb{B}$. In the exercises we develop the invariant Laplacian, Green's function, and invariant measure on $\mathbb{H}$.

A real-valued C^2 function f on $\mathbb{B}$ is defined to be either $\mathcal{H}$-harmonic or $\mathcal{H}$-subharmonic on $\mathbb{B}$ depending on whether $\Delta_h f = 0$ or $\Delta_h f \geq 0$. It is well known that a continuous function f is harmonic in the unit disc $\mathbb{D}$ if and only if for all r, $0 < r < 1$, and $w \in \mathbb{D}$,

$$f(w) = \frac{1}{2\pi} \int_0^{2\pi} f(\varphi_w(re^{it}))\, dt, \tag{3}$$

where φ_w is the Möbius transformation of $\mathbb{D}$ given by (2). The above is called the **invariant mean-value property**. One of the first results proved in Chapter 4 is the following analogue of the invariant mean-value property: A real-valued C^2 function f is $\mathcal{H}$-subharmonic on $\mathbb{B}$ if and only if for all $a \in \mathbb{B}$ and $0 < r < 1$,

$$f(a) \leq \int_{\mathbb{S}} f(\varphi_a(rt)) d\sigma(t), \tag{4}$$

with equality if and only if f is $\mathcal{H}$-harmonic on $\mathbb{B}$. In the above, $\mathbb{S}$ is the unit sphere in $\mathbb{R}^n$, σ is normalized surface measure on $\mathbb{S}$, and φ_a is the Möbius transformation of $\mathbb{B}$ mapping 0 to a with $\varphi_a(\varphi_a(x)) = x$. The integral in (4) is an average of f over the hyperbolic or non-Euclidean sphere $\{\varphi_a(rt) : t \in \mathbb{S}\}$ whose hyperbolic center is a. Inequality (4) is then used in Section 4.3 to extend the definition of $\mathcal{H}$-subharmonic to the class of upper semicontinuous functions on $\mathbb{B}$. The remainder of the chapter is devoted to extending some of the standard results about subharmonic functions to $\mathcal{H}$-subharmonic functions on $\mathbb{B}$. We conclude the chapter with a discussion of quasi-nearly $\mathcal{H}$-subharmonic functions and prove several inequalities involving these functions that will prove useful later in the text.

The Poisson kernel P_h for Δ_h is introduced in Chapter 5. In Section 5.1 we prove using Green's formula that for $(a, t) \in \mathbb{B} \times \mathbb{S}$,

$$P_h(a, t) = -\lim_{r \to 1} nr^{n-1}(1 - r^2)^{2-n} \langle \nabla G_a(rt), t \rangle,$$

where $G_a(rt) = G_h(a, rt)$ is the Green's function for Δ_h. This immediately gives

$$P_h(x, t) = \left(\frac{1 - |x|^2}{|x - t|^2} \right)^{n-1}, \qquad (x, t) \in \mathbb{B} \times \mathbb{S}.$$

The standard results for Poisson integrals of continuous functions are included in Section 5.3, and in Section 5.2 we prove a result of P. Jaming [43] that provides an integral representation of the Euclidean Poisson kernel in terms of the hyperbolic Poisson kernel. In Section 5.5 we investigate the eigenfunctions of Δ_h. We close the section with a brief discussion of the Poisson kernel on $\mathbb{H}$.

In Chapter 6 we consider the spherical harmonic expansions of $\mathcal{H}$-harmonic functions. One of the key results of this section is that if p_α is a spherical harmonic of degree α on $\mathbb{S}$, then the Poisson integral $P_h[p_\alpha]$ of p_α is given by

$$P_h[p_\alpha](x) = |x|^\alpha S_{n,\alpha}(|x|)p_\alpha\left(\tfrac{x}{|x|}\right),$$

where $S_{n,\alpha}$ is given by a hypergeometric function. Interestingly, when n is even, $S_{n,\alpha}(r)$ is simply a polynomial in r of degree $n-2$. These results are then used to show how the Poisson integral $P_h[q]$ can be computed for any polynomial q on $\mathbb{S}$. As an example, in $\mathbb{R}^4$, the $\mathcal{H}$-harmonic function with boundary values t_1^2 is given by $P_h[t_1^2](x) = \frac{1}{4} + (2-|x|^2)(x_1^2 - \frac{1}{4}|x|^2)$. In contrast, the Euclidean harmonic function h with boundary values t_1^2 is given by $h(x) = \frac{1}{4}(1-|x|^2) + x_1^2$. Finally, in Section 6.3 we follow the methods of P. Ahern, J. Bruna, and C. Cascante [2] to derive the spherical harmonic expansion of $\mathcal{H}$-harmonic functions on $\mathbb{B}$.

Chapter 7 is devoted to the study of Hardy and Hardy–Orlicz type spaces of $\mathcal{H}$-harmonic and $\mathcal{H}$-subharmonic functions on $\mathbb{B}$. In Chapter 8, we study the boundary behavior of Poisson integrals on $\mathbb{B}$. This chapter contains many of the standard results concerning non-tangential and radial maximal functions. In addition to proving the usual Fatou theorem (Theorem 8.3.3) concerning non-tangential limits of Poisson integrals of measures, we also include a proof of a local Fatou theorem of I. Privalov [68] for $\mathcal{H}$-harmonic functions on $\mathbb{B}$.

The Riesz decomposition theorem for $\mathcal{H}$-subharmonic functions is proved in Chapter 9. The main result of this chapter (Corollary 9.1.3) proves that if f is $\mathcal{H}$-subharmonic on $\mathbb{B}$ and f has an $\mathcal{H}$-harmonic majorant, then

$$f(x) = F_f(x) - \int_{\mathbb{B}} G_h(x,y)\,d\mu_f(y),$$

where μ_f is the Riesz measure of f and F_f is the least $\mathcal{H}$-harmonic majorant of f. In Section 9.2 we include several applications of the Riesz decomposition theorem, including a Hardy–Stein identity for non-negative $\mathcal{H}$-subharmonic functions for which f^p, $p \ge 1$, has an $\mathcal{H}$-harmonic majorant on $\mathbb{B}$. In Section 9.3 we extend a result of D. H. Armitage [8] concerning the integrability of non-negative superharmonic functions. We conclude the chapter by proving that invariant Green potentials of measures have radial limit zero almost everywhere on $\mathbb{S}$, and provide an example of a measure μ for which the Green potential of μ has non-tangential limit $+\infty$ almost everywhere on $\mathbb{S}$.

Finally, in Chapter 10 we introduce and investigate basic properties of weighted Bergman- and Dirichlet-type spaces of $\mathcal{H}$-harmonic functions on $\mathbb{B}$, denoted respectively by $\mathcal{B}_\gamma^p$ and $\mathcal{D}_\gamma^p$. These spaces consist of the set of $\mathcal{H}$-harmonic functions f on $\mathbb{B}$ for which f, respectively $|\nabla^h f|$, are in

$L^p((1-|x|^2)^\gamma d\tau(x))$, $0 < p < \infty$, $\gamma > 0$, where τ is the invariant measure on $\mathbb{B}$ and ∇^h is the invariant gradient on $\mathbb{B}$. One of the main results of this chapter is that if $\gamma > (n-1)$, then $f \in \mathcal{B}^p_\gamma$ if and only if $f \in \mathcal{D}^p_\gamma$ for all p, $0 < p < \infty$. In Section 10.4 we investigate the integrability of functions in $\mathcal{B}^p_\gamma$ and $\mathcal{D}^p_\gamma$ for $\gamma \le (n-1)$. This chapter also contains a discussion of Möbius invariant spaces of $\mathcal{H}$-harmonic functions and the Berezin transform on $\mathbb{B}$. We conclude the chapter with three theorems of Hardy and Littlewood for $\mathcal{H}$-harmonic functions, and the Littlewood–Paley inequalities for $\mathcal{H}$-subharmonic functions.

At the end of each chapter, I have included a set of exercises dealing with the topics discussed. Many of these problems involve routine computations and inequalities not included in the text. They also provide examples relevant to the topics of the chapter. Also included are problems whose solutions may be suitable for possible publication. The latter are marked with an asterisk.

Acknowledgments

The preliminary draft of these notes comprising most of Chapters 1–3 and parts of Chapters 5 and 6 was written while the author was on sabbatical leave at the CRM at the University of Montreal during fall 1999. Since 1999 there has been considerable research activity on hyperbolic function theory on $\mathbb{B}$. Included in this has been the work of P. Jaming [41], [42], [43], S. Grellier and P. Jaming [31], M. Jevtić [44], and the author [89], among others.

I would like to thank my good friend Paul Gauthier at the University of Montreal for providing the motivation for this project. His many questions about harmonic functions on real hyperbolic space, to which I did not know the answers, encouraged me to learn more about this interesting subject. I am also indebted to Paul for using the preliminary draft of these notes in one of his courses and for pointing out a number of errors. I would also like to express my appreciation to John Taylor at McGill for providing me with a copy of his notes on rank one symmetric spaces, and to thank the Director and Staff at the CRM for their hospitality during my visit. Finally, I would like to thank Wolfgang Woess Universität Graz, for bringing my original manuscript to the attention of Sam Harrison, Editor at Cambridge University Press, and the staff at Cambridge University Press and SPi Global for their assistance in the completion of the project.

Manfred Stoll
Columbia, SC
stoll@math.sc.edu

1

Möbius Transformations

In this chapter we provide a brief review of Möbius transformations on n-dimensional Euclidean space $\mathbb{R}^n$ ($n \geq 2$). A good reference for these topics is the monograph by A. F. Beardon [11]. First, however, we begin with a review of notation that will be used throughout these notes.

1.1 Notation

For $x, y \in \mathbb{R}^n$ we let $\langle x, y \rangle = \sum_{j=1}^{n} x_j y_j$ denote the usual inner product on $\mathbb{R}^n$ and $|x| = \sqrt{\langle x, x \rangle}$ the length of the vector x. For $a \in \mathbb{R}^n$ and $r > 0$, the ball $B(a, r)$ and sphere $S(a, r)$ are given respectively by

$$B(a, r) = \{x \in \mathbb{R}^n : |x - a| < r\},$$
$$S(a, r) = \{x \in \mathbb{R}^n : |x - a| = r\}.$$

The unit ball and unit sphere with center at the origin will simply be denoted by $\mathbb{B}$ and $\mathbb{S}$ respectively.[1] The **one-point compactification** of $\mathbb{R}^n$, denoted $\hat{\mathbb{R}}^n$, is obtained by appending the point ∞ to $\mathbb{R}^n$. A subset U of $\hat{\mathbb{R}}^n = \mathbb{R}^n \cup \{\infty\}$ is open if it is an open subset of $\mathbb{R}^n$, or if U is the complement in $\hat{\mathbb{R}}^n$ of a compact subset C of $\mathbb{R}^n$. With this topology $\hat{\mathbb{R}}^n$ is compact.

For a subset D of $\mathbb{R}^n$, $\overline{D}$ denotes the closure of D, Int (D) the interior of D, ∂D the boundary of D, and $\widetilde{D}$ the complement of D in $\mathbb{R}^n$. Also if E and F are sets, $E \setminus F$ denotes the complement of F in E, that is, $E \setminus F = E \cap \widetilde{F}$.

The study of functions of n-variables is simplified with the use of multi-index notation. For an ordered n-tuple $\alpha = (\alpha_1, \ldots, \alpha_n)$, where each α_j is a non-negative integer, the following notational conventions will be used throughout:

[1] If we wish to emphasize the dimension n, we will use the notation $\mathbb{B}_n$ and $\mathbb{S}_n$ to denote the unit ball and sphere in $\mathbb{R}^n$.

$$|\alpha| = \alpha_1 + \cdots + \alpha_n, \quad \alpha! = \alpha_1! \cdots \alpha_n!, \quad x^\alpha = x_1^{\alpha_1} \cdots x_n^{\alpha_n},$$

and

$$D^\alpha f = \frac{\partial^{|\alpha|} f}{\partial x_1^{\alpha_1} \cdots \partial x_n^{\alpha_n}}.$$

If Ω is an open subset of $\mathbb{R}^n$, we denote by $C^k(\Omega)$, $k = 0, 1, 2, \ldots$ the set of real-valued (or complex-valued) functions f on Ω for which $D^\alpha f$ exists and is continuous for all multi-indices α with $|\alpha| \le k$. Thus $C^0(\Omega)$, or simply $C(\Omega)$, denotes the set of real-valued (or complex-valued) continuous functions on Ω, and $C^\infty(\Omega)$ the set of infinitely differentiable functions on Ω. Also, the set of functions $f \in C^k(\Omega)$ for which $D^\alpha f$, $|\alpha| \le k$, has a continuous extension to $\overline{\Omega}$ will be denoted by $C^k(\overline{\Omega})$. If $f : \Omega \mapsto \mathbb{R}$, then the **support** of f, denoted $\operatorname{supp} f$, is defined as

$$\operatorname{supp} f = \overline{\{x \in \Omega : f(x) \neq 0\}}.$$

The set of continuous functions on Ω with compact support will be denoted by $C_c(\Omega)$. The notations $C_c^k(\Omega)$ and $C_c^\infty(\Omega)$ have the obvious meanings.

A linear transformation $A : \mathbb{R}^n \mapsto \mathbb{R}^n$ is said to be **orthogonal** if $|Ax| = |x|$ for all $x \in \mathbb{R}^n$. The set of orthogonal transformations of $\mathbb{R}^n$ will be denoted by $O(n)$. If A is represented by the $n \times n$ matrix $(a_{i,j})$, then A is orthogonal if and only if

$$\sum_{k=1}^n a_{i,k} a_{j,k} = \delta_{i,j} = \begin{cases} 1 & i = j, \\ 0, & i \neq j. \end{cases}$$

If $\psi(x) = (\psi_1(x), \ldots, \psi_n(x))$ is a C^1 mapping of an open subset Ω of $\mathbb{R}^n$ into $\mathbb{R}^n$, then the derivative $\psi'(x)$ is the $n \times n$ matrix given by

$$\psi'(x) = \left(\frac{\partial \psi_i}{\partial x_j} \right)_{i,j=1}^n,$$

and the **Jacobian** J_ψ of the transformation ψ is given by $J_\psi(x) = \det \psi'(x)$.

1.2 Inversion in Spheres and Planes

Definition 1.2.1 *The* ***inversion***[2] *(or reflection) in the sphere* $S(a, r)$ *is the function* $\phi(x)$ *defined by*

[2] Although we will mainly be interested in the case $n \ge 2$, the formulas for inversions in spheres and planes are still meaningful when $n = 1$.

$$\phi(x) = a + \left(\frac{r}{|x-a|}\right)^2 (x-a). \tag{1.2.1}$$

The inversion in the unit sphere $\mathbb{S}$ is the mapping $\phi(x) = x^*$ where

$$x^* = \begin{cases} \dfrac{x}{|x|^2} & x \neq 0, \infty, \\ 0 & x = \infty, \\ \infty & x = 0. \end{cases}$$

Thus (1.2.1) can now be rewritten as

$$\phi(x) = a + r^2(x-a)^*.$$

The reflection $\phi(x)$ is not defined at $x = a$. Since $|\phi(x)| \to \infty$ as $x \to a$ we set $\phi(a) = \infty$. Also, since $\lim_{|x|\to\infty} |\phi(x) - a| = 0$, we set $\phi(\infty) = a$. Thus ϕ is defined on all of $\hat{\mathbb{R}}^n$, and it is easily shown that ϕ is continuous in the topology of $\hat{\mathbb{R}}^n$. A straightforward computation also shows that $\phi(\phi(x)) = x$ for all $x \in \hat{\mathbb{R}}^n$. Thus ϕ is a one-to-one continuous map of $\hat{\mathbb{R}}^n$ onto $\hat{\mathbb{R}}^n$ satisfying $\phi(x) = x$ if and only $x \in S(a, r)$.

In addition to reflection in a sphere we also have reflection in a plane. For $a \in \mathbb{R}^n$, $a \neq 0$, and $t \in \mathbb{R}$, the plane $P(a, t)$ is defined by

$$P(a,t) = \{x \in \mathbb{R}^n : \langle x, a\rangle = t\}.$$

By convention ∞ belongs to every plane $P(a, t)$.

Definition 1.2.2 *The* **inversion** *(or reflection) in the plane $P(a,t)$ is the function $\psi(x)$ defined by*

$$\psi(x) = x + \lambda a,$$

where $\lambda \in \mathbb{R}$ is chosen so that $\frac{1}{2}(x + \psi(x)) \in P(a,t)$.

Solving for λ gives

$$\psi(x) = x - 2[\langle x, a\rangle - t]a^*, \quad x \in \mathbb{R}^n. \tag{1.2.2}$$

For the mapping ψ we have

$$|\psi(x)|^2 = |x|^2 + O(|x|),$$

and as a consequence $\lim_{|x|\to\infty} |\psi(x)| = \infty$. Thus as above we define $\psi(\infty) = \infty$. With this definition the mapping ψ again satisfies $\psi(\psi(x)) = x$ for all $x \in \hat{\mathbb{R}}^n$. Thus ψ is a one-to-one continuous map of $\hat{\mathbb{R}}^n$ onto itself with $\psi(x) = x$ if and only if $x \in P(a,t)$. It is well known that each inversion (in a sphere or a plane) is orientation-reversing and conformal (see [11, Theorem 3.1.6]).

1.3 Möbius Transformations

Definition 1.3.1 *A* ***Möbius transformation*** *of* $\hat{\mathbb{R}}^n$ *is a finite composition of inversions in spheres or planes.*

Clearly the composition of two Möbius transformations is again a Möbius transformation, as is the inverse of a Möbius transformation. The group of Möbius transformations on $\hat{\mathbb{R}}^n$ is called the **general Möbius group** and is denoted by GM $(\hat{\mathbb{R}}^n)$. Although not immediately obvious, both translation and magnification by a constant are Möbius transformations. The translation $x \mapsto x + a$, $a \in \mathbb{R}^n$, is the composition of inversion in the plane $\langle x, a\rangle = 0$ followed by inversion in the plane $\langle x, a\rangle = \frac{1}{2}|a|^2$. Likewise, the magnification or scalar multiplication $x \mapsto kx$, $k > 0$, is also a Möbius transformation in that it is the inversion in $\mathbb{S}$ followed by the inversion in $S(0, \sqrt{k})$. Furthermore, every Euclidean isometry of $\mathbb{R}^n$ is a composition of at most $n + 1$ reflections in planes ([11, Theorem 3.1.3]).

We conclude this section by showing that every Möbius transformation maps a sphere or plane onto a sphere or plane. We will use the term "sphere" to denote either a sphere of the form $S(a, r)$ or a plane $P(a, t)$. Since every inversion ψ in a plane $P(a, t)$ can be written as

$$\psi(x) = x + \lambda a,$$

the mapping ψ clearly maps a "sphere" onto a "sphere." To show that an inversion ϕ in a sphere $S(a, r)$ preserves "spheres," it suffices to show that the mapping x^* preserves "spheres."

For any set $E \subset \mathbb{R}^n$, we let $E^* = \{x^* : x \in E\}$. A set $E \subset \mathbb{R}^n$ is a sphere or a plane if and only if

$$E = \{x \in \mathbb{R}^n : b|x|^2 - 2\langle x, a\rangle + c = 0\},$$

where b and c are real and $a \in \mathbb{R}^n$. By convention, ∞ satisfies this equation if and only if $b = 0$, that is, E is a plane. Now it is easily seen that E^* has the same form with the roles of b and c reversed. Finally, it is an easy exercise to show that for $a \in \mathbb{R}^n$ and $r > 0$,

$$S^*(a, r) = \begin{cases} S\left(\dfrac{a}{(|a|^2 - r^2)}, \dfrac{r}{||a|^2 - r^2|}\right) & \text{if } 0 \notin S(a, r), \\ P(a, \frac{1}{2}) & \text{if } 0 \in S(a, r). \end{cases} \tag{1.3.1}$$

We conclude this section with one more useful formula that will be required later. If ϕ is inversion in the sphere $S(a, r)$, then a straightforward computation gives

$$|\phi(y) - \phi(x)| = \frac{r^2|y - x|}{|x - a||y - a|}. \tag{1.3.2}$$

For details on the above the reader is referred to [11].

2

Möbius Self-Maps of the Unit Ball

In this chapter we will provide a characterization of the Möbius transformations of $\hat{\mathbb{R}}^n$ mapping the unit ball $\mathbb{B}$ onto $\mathbb{B}$ that is similar to the characterization of the Möbius mappings of the unit disc in $\mathbb{C}$ onto itself.

In the complex plane $\mathbb{C}$, every analytic Möbius transformation ψ mapping the unit disc $\mathbb{D}$ onto itself can be written as $\psi(z) = e^{i\theta}\varphi_w(z)$, where for $w \in \mathbb{D}$,

$$\varphi_w(z) = \frac{w - z}{1 - \overline{w}z}.$$

The mappings $\varphi_w(z)$ satisfy $\varphi_w(0) = w$, $\varphi_w(w) = 0$, and $\varphi_w(\varphi_w(z)) = z$ for all $z \in \mathbb{D}$. Furthermore, the mapping $\varphi_w(z)$ also satisfies

$$1 - |\varphi_w(z)|^2 = \frac{(1 - |z|^2)(1 - |w|^2)}{|1 - \overline{w}z|^2}.$$

2.1 Möbius Transformations of $\mathbb{B}$

In this section we define an analogous family of Möbius transformations $\{\varphi_a : a \in \mathbb{B}\}$ mapping $\mathbb{B}$ onto $\mathbb{B}$ having the property that every Möbius transformation ψ mapping $\mathbb{B}$ onto itself can be written as $\psi = A \circ \varphi_a$, where $a \in \mathbb{B}$ and $A \in O(n)$. For $a \in \mathbb{B}$, we first set

$$\psi_a(x) = a + (1 - |a|^2)(a - x)^*. \tag{2.1.1}$$

Since the mapping ψ_a is a composition of Möbius transformations, ψ_a is a Möbius transformation of $\hat{\mathbb{R}}^n$ mapping 0 to a^* and a to ∞. By a straightforward computation we have

$$|\psi_a(x)|^2 = \frac{|a - x|^2 + (1 - |a|^2)(1 - |x|^2)}{|a - x|^2}, \tag{2.1.2}$$

and as a consequence

$$|\psi_a(x)|^2 - 1 = \frac{(1-|a|^2)(1-|x|^2)}{|x-a|^2}. \tag{2.1.3}$$

From the above it follows immediately that ψ_a maps $\mathbb{B}$ onto $\hat{\mathbb{R}}^n \setminus \overline{\mathbb{B}}$.

We now define the mapping φ_a by

$$\varphi_a(x) = \psi_a(x)^* = \frac{\psi_a(x)}{|\psi_a(x)|^2}. \tag{2.1.4}$$

If we set[1]

$$\rho(x,a) = |x-a|^2 + (1-|a|^2)(1-|x|^2) = |a|^2|a^* - x|^2, \tag{2.1.5}$$

then the mapping φ_a can be expressed as

$$\varphi_a(x) = \frac{a|x-a|^2 + (1-|a|^2)(a-x)}{\rho(x,a)}. \tag{2.1.6}$$

As a consequence of (2.1.3)

$$1 - |\varphi_a(x)|^2 = \frac{(1-|x|^2)(1-|a|^2)}{\rho(x,a)}. \tag{2.1.7}$$

Thus φ_a is a Möbius transformation mapping $\mathbb{B}$ onto $\mathbb{B}$ with $\varphi_a(0) = a$ and $\varphi_a(a) = 0$. That φ_a maps $\mathbb{B}$ onto $\mathbb{B}$ follows immediately from the fact that ψ_a maps $\mathbb{B}$ onto $\hat{\mathbb{R}}^n \setminus \overline{\mathbb{B}}$ and that x^* maps $\hat{\mathbb{R}}^n \setminus \overline{\mathbb{B}}$ onto $\mathbb{B}$. We will shortly prove that φ_a also satisfies $\varphi_a(\varphi_a(x)) = x$ for all $x \in \mathbb{B}$. In the unit disk $\mathbb{D}$, for $z, w \in \mathbb{D}$, $\rho(z,w) = |1 - \overline{w}z|^2$ and the mappings $\varphi_w(z)$ as defined by (2.1.6) are precisely the functions $(w-z)/(1-\overline{w}z)$.

One of the advantages of the mappings φ_a is that the function $(a,x) \mapsto \varphi_a(x)$ is not only continuous on $\overline{\mathbb{B}} \times \overline{\mathbb{B}}$ but also differentiable in each of the variables. At this point we will include several computations involving derivatives of the mappings φ_a that will be required in the proof of Theorem 2.1.2 and also later in the sequel. Let $y_j(x)$ denote the jth coordinate of $y(x) = \varphi_a(x)$. Then by straightforward computations we have

$$\frac{\partial y_j}{\partial x_i}(0) = -\delta_{i,j}(1-|a|^2), \qquad \frac{\partial y_j}{\partial x_i}(a) = \frac{-\delta_{i,j}}{(1-|a|^2)}, \tag{2.1.8}$$

$$\frac{\partial^2 y_j}{\partial x_i^2}(0) = (1-|a|^2)[2a_j - 4a_i\delta_{i,j}]. \tag{2.1.9}$$

Hence

$$\varphi_a'(0) = -(1-|a|^2)\,I \quad \text{and} \quad \varphi_a'(a) = -(1-|a|^2)^{-1}I,$$

where I is the $n \times n$ identity matrix.

[1] In [11] the function $\sqrt{\rho(x,a)}$ is denoted by $[x,a]$.

Since the following theorem is well known, we state it without proof. A proof may be found in [11, Theorem 3.4.1].

Theorem 2.1.1 *Let ψ be a Möbius transformation of $\hat{\mathbb{R}}^n$ satisfying $\psi(0) = 0$ and $\psi(\mathbb{B}) = \mathbb{B}$. Then $\psi(x) = Ax$ for some orthogonal transformation A.*

We denote by $\mathcal{M}(\mathbb{B})$ the set of all **Möbius transformations of $\mathbb{B}$ onto $\mathbb{B}$**. It is an immediate consequence of the following theorem that the set $\mathcal{M}(\mathbb{B})$ forms a group called the Möbius group of $\mathbb{B}$.

Theorem 2.1.2 *For $a \in \mathbb{B}$, let φ_a be defined by (2.1.6). Then*

(a) *φ_a is a one-to-one Möbius mapping of $\mathbb{B}$ onto $\mathbb{B}$ satisfying*

$$\varphi_a(0) = a, \quad \varphi_a(a) = 0, \quad \text{and} \quad \varphi_a(\varphi_a(x)) = x$$

for all $x \in \mathbb{B}$.

(b) *If $\psi \in \mathcal{M}(\mathbb{B})$, then there exists an orthogonal transformation A and $a \in \mathbb{B}$ such that $\psi(x) = A\varphi_a(x)$.*

Proof. To prove (a) it only remains to be shown that $\varphi_a(\varphi_a(x)) = x$ for all $x \in \mathbb{B}$. Set $\psi(x) = (\varphi_a \circ \varphi_a)(x)$. Then ψ is a Möbius transformation of $\hat{\mathbb{R}}^n$ mapping $\mathbb{B}$ onto $\mathbb{B}$ satisfying $\psi(0) = 0$. Thus $\psi(x) = Ax$ for some orthogonal transformation A. But then $A = \psi'(0)$. On the other hand, by the chain rule and Equations (2.1.8)

$$\psi'(0) = \varphi_a'(a)\varphi_a'(0) = I.$$

Hence $A = I$ and thus $\varphi_a(\varphi_a(x)) = x$ for all $x \in \mathbb{B}$.

(b) Let $\psi \in \mathcal{M}(\mathbb{B})$ and let $a = \psi^{-1}(0)$. Then $\psi \circ \varphi_a$ is a Möbius transformation of $\mathbb{B}$ that fixes the origin. Thus $\psi \circ \varphi_a(x) = Ax$ for some orthogonal transformation A. But then by (a) we have $\psi(x) = A\varphi_a(x)$. □

Prior to introducing the hyperbolic metric on $\mathbb{B}$ we prove an identity for mappings $\psi \in \mathcal{M}(\mathbb{B})$.

Theorem 2.1.3 *If $\psi \in \mathcal{M}(\mathbb{B})$, then for all $x, y \in \mathbb{B}$,*

$$\frac{|\psi(x) - \psi(y)|^2}{(1 - |\psi(x)|^2)(1 - |\psi(y)|^2)} = \frac{|x - y|^2}{(1 - |x|^2)(1 - |y|^2)}.$$

Proof. Although this identity could be proved using the mappings φ_a, it appears to be easier to use the mappings σ_a defined as follows: for $a \in \mathbb{B}$, $a \neq 0$, let σ_a denote the inversion in the sphere $S(a^*, \sqrt{|a^*|^2 - 1})$, that is,

$$\sigma_a(x) = a^* + (|a^*|^2 - 1)(x - a^*)^*. \tag{2.1.10}$$

Then $\sigma_a(0) = a$, $\sigma_a(a) = 0$, and since σ_a is an inversion, $\sigma_a(\sigma_a(x)) = x$ for all $x \in \hat{\mathbb{R}}^n$. Also, by identity (1.3.2),

$$|\sigma_a(x)|^2 = |\sigma_a(x) - \sigma_a(a)|^2 = \frac{(|a^*|^2 - 1)^2|x - a|^2}{|x - a^*|^2|a - a^*|^2},$$

which upon simplification gives

$$|\sigma_a(x)|^2 = \frac{|x - a|^2}{\rho(x, a)}.$$

Thus

$$1 - |\sigma_a(x)|^2 = \frac{(1 - |x|^2)(1 - |a|^2)}{\rho(x, a)}. \tag{2.1.11}$$

Hence $\sigma_a \in \mathcal{M}(\mathbb{B})$.[2] Again by (1.3.2) we obtain

$$|\sigma_a(x) - \sigma_a(y)|^2 = \frac{(1 - |a|^2)^2|x - y|^2}{\rho(x, a)\rho(y, a)}.$$

Combining this with (2.1.11) now gives

$$\frac{|\sigma_a(x) - \sigma_a(y)|^2}{(1 - |\sigma_a(x)|^2)(1 - |\sigma_a(y)|^2)} = \frac{|x - y|^2}{(1 - |x|^2)(1 - |y|^2)}.$$

Finally, as in the proof of Theorem 2.1.2(b), every $\psi \in \mathcal{M}(\mathbb{B})$ can be expressed as $\psi(x) = A\sigma_a(x)$ for some $A \in O(n)$ and $a \in \mathbb{B}$. From this the result now follows. □

As a consequence of the identity in Theorem 2.1.3, for $\psi \in \mathcal{M}(\mathbb{B})$,

$$\lim_{y \to x} \frac{|\psi(y) - \psi(x)|}{|y - x|} = \frac{1 - |\psi(x)|^2}{1 - |x|^2}. \tag{2.1.12}$$

This result will be required in proving the $\mathcal{M}$-invariance of the hyperbolic metric on $\mathbb{B}$.

2.2 The Hyperbolic Metric on $\mathbb{B}$

The element of arclength ds for the **hyperbolic metric** d_h on $\mathbb{B}$ is given by

$$ds = \frac{2|dx|}{1 - |x|^2}. \tag{2.2.1}$$

Thus if $\gamma : [0, 1] \mapsto \mathbb{B}$ is a C^1 curve in $\mathbb{B}$, the **hyperbolic length** $L(\gamma)$ of γ is given by

[2] Even though the mappings σ_a are easier to work with, they have the disadvantage that $\lim_{a \to 0} \sigma_a(x)$ does not exist.

$$L(\gamma) = \int_0^1 \frac{2|\gamma'(t)|\,dt}{1-|\gamma(t)|^2},$$

and for $a, b \in \mathbb{B}$, the **hyperbolic distance** $d_h(a,b)$ between a and b is defined by

$$d_h(a,b) = \inf_{\gamma} L(\gamma),$$

where the infimum is taken over all C^1 curves $\gamma : [0,1] \mapsto \mathbb{B}$ with $\gamma(0) = a$ and $\gamma(1) = b$. From this we immediately obtain that for $x \in \mathbb{B}$,

$$d_h(0,x) = \log\left(\frac{1+|x|}{1-|x|}\right). \tag{2.2.2}$$

Theorem 2.2.1 *For all $\psi \in \mathcal{M}(\mathbb{B})$ and $a, b \in \mathbb{B}$, $d_h(\psi(a), \psi(b)) = d_h(a,b)$.*

Proof. To prove the theorem it suffices to prove that $L(\psi \circ \gamma) = L(\gamma)$ for all C^1 curves γ and $\psi \in \mathcal{M}(\mathbb{B})$. If we set $\sigma(t) = \psi(\gamma(t))$, then σ is a C^1 curve and

$$\begin{aligned}
|\sigma'(t)| &= \lim_{h\to 0}\left|\frac{\sigma(t+h)-\sigma(t)}{h}\right| \\
&= \lim_{h\to 0}\frac{|\psi(\gamma(t+h)) - \psi(\gamma(t))|}{|h|},
\end{aligned}$$

which by (2.1.12)

$$= |\gamma'(t)|\left(\frac{1-|\psi(\gamma(t))|^2}{1-|\gamma(t)|^2}\right).$$

Thus

$$\frac{|\sigma'(t)|}{1-|\sigma(t)|^2} = \frac{|\gamma'(t)|}{1-|\gamma(t)|^2}.$$

From this it now follows that $L(\sigma) = L(\gamma)$, thus proving the claim. □

As a consequence of (2.2.2) and Theorem 2.2.1, for $a, b \in \mathbb{B}$,

$$d_h(a,b) = d_h(0, \varphi_a(b)) = \log\left(\frac{1+|\varphi_a(b)|}{1-|\varphi_a(b)|}\right). \tag{2.2.3}$$

Some brief computations also give

$$\sinh^2 \tfrac{1}{2} d_h(a,b) = \frac{|a-b|^2}{(1-|a|^2)(1-|b|^2)}, \tag{2.2.4}$$

or

$$d_h(a,b) = 2\,\text{Arcsinh}\left[\frac{|a-b|}{\sqrt{(1-|a|^2)(1-|b|^2)}}\right]$$
$$= 2\,\text{Arctanh}\left[\frac{|a-b|}{\sqrt{\rho(a,b)}}\right],$$

where $\rho(a,b)$ is defined by (2.1.5).

For $0 < r < 1$ we will denote $B(0,r)$ and $S(0,r)$ by B_r and S_r respectively. As in [72, p. 29], for $a \in \mathbb{B}$ and $0 < r < 1$, we let $E(a,r) = \varphi_a(B_r)$. Since φ_a is an involution,

$$E(a,r) = \{x \in \mathbb{B} : |\varphi_a(x)| < r\} = \left\{x \in \mathbb{B} : d_h(a,x) < \log\left(\tfrac{1+r}{1-r}\right)\right\}. \quad (2.2.5)$$

Thus $E(a,r)$ is a **hyperbolic ball** with hyperbolic center a and hyperbolic radius

$$\rho = \log\left(\frac{1+r}{1-r}\right) = 2\,\text{Arctanh}\, r.$$

However, $E(a,r)$ is also a Euclidean ball whose center and radius are given in the following theorem.

Theorem 2.2.2 *For $a \in \mathbb{B}$ and $0 < r < 1$, $E(a,r) = B(c_a, \rho_a)$ where*

$$c_a = \frac{(1-r^2)a}{(1-|a|^2r^2)} \quad \textit{and} \quad \rho_a = \frac{r(1-|a|^2)}{(1-|a|^2r^2)}.$$

Proof. To prove the result we first determine the image of S_r under the mapping φ_a. Let ψ_a be the mapping defined by (2.1.1). If $|a| \neq r$, then by (1.3.1)

$$\psi_a(S_r) = S\left(\frac{(1-r^2)a}{(|a|^2-r^2)}, \frac{(1-|a|^2)r}{||a|^2-r^2|}\right).$$

Since $\varphi_a = \psi_a^*$, using (1.3.1) again gives

$$\varphi_a(S_r) = S\left(\frac{(1-r^2)a}{(1-r^2|a|^2)}, \frac{(1-|a|^2)r}{(1-r^2|a|^2)}\right). \quad (2.2.6)$$

On the other hand, if $|a| = r$, $\psi_a(S_r) = P(a, \frac{1}{2}(1+|a|^2))$. Taking the inversion of $\psi_a(S_r)$ gives

$$\varphi_a(S_r) = \{x : (1+|a|^2)|x|^2 - 2\langle a,x\rangle = 0\}.$$

This, however, is simply the equation of the sphere given in (2.2.6) with $r = |a|$. Since φ_a is continuous and $a \in B(c_a, \rho_a)$, $\varphi_a(B_r) \subset B(c_a, \rho_a)$. Finally, since φ_a is an involution, $\varphi_a(B_r) = B(c_a, \rho_a)$. □

We conclude this section with the **pseudo-hyperbolic metric** on $\mathbb{B}$. For $a,\ b \in \mathbb{B}$ set

$$d_{ph}(a,b) = |\varphi_a(b)|. \tag{2.2.7}$$

Theorem 2.2.3 *d_{ph} is a metric on $\mathbb{B}$.*

Proof. Since φ_a is one-to-one the only fact that needs to be proved is that d_{ph} satisfies the triangle inequality, that is,

$$|\varphi_a(b)| \le |\varphi_a(x)| + |\varphi_b(x)|$$

for all $a,\ b,\ x \in \mathbb{B}$. We first note that $d_{ph}(a,b) = f(d_h(a,b))$ where

$$f(t) = \frac{e^t - 1}{e^t + 1}, \quad t \in [0,\infty).$$

Since f is increasing on $[0,\infty)$, we have

$$d_{ph}(a,b) = f(d_h(a,b)) \le f(d_h(a,x) + d_h(x,b)).$$

Finally, since

$$f(x+y) \le f(x) + f(y) \quad \text{for all} \quad x,y \in [0,\infty), \tag{2.2.8}$$

we have

$$f(d_h(a,x) + d_h(x,b)) \le f(d_h(a,x)) + f(d_h(x,b)) = d_{ph}(a,x) + d_{ph}(x,b),$$

which proves the result. For the proof of (2.2.8) we first note that

$$\frac{e^x - 1}{e^x + 1} + \frac{e^y - 1}{e^y + 1} = \frac{2(e^{x+y} - 1)}{(e^x + 1)(e^y + 1)}.$$

Since $e^{x+y} - e^x - e^y + 1 = (e^x - 1)(e^y - 1) \ge 0$, we have

$$2(e^{x+y} + 1) - (e^x + 1)(e^y + 1) \ge 0.$$

Therefore

$$\frac{2}{(e^x + 1)(e^y + 1)} \ge \frac{1}{e^{x+y} + 1}$$

which proves (2.2.8) and thus the result. □

2.3 Hyperbolic Half-Space $\mathbb{H}$

In this final section we briefly consider hyperbolic half-space in $\mathbb{R}^n$.

Definition 2.3.1 *For $n \geq 2$, the* **upper half-space** $\mathbb{H}$ *or* $\mathbb{H}_n$ *in* $\mathbb{R}^n$ *is defined by*

$$\mathbb{H} = \{x \in \mathbb{R}^n : x_n > 0\}.$$

For each $x \in \mathbb{R}^{n-1}$, let $\tilde{x} \in \mathbb{R}^n$ be defined by

$$\tilde{x} = (x, 0) = (x_1, \ldots, x_{n-1}, 0).$$

For each inversion ϕ on $\hat{\mathbb{R}}^{n-1}$, we define an inversion $\tilde{\phi}$ acting on $\hat{\mathbb{R}}^n$ as follows. If ϕ is an inversion in $S(a, r)$, then $\tilde{\phi}$ is the inversion in $S(\tilde{a}, r)$; if ϕ is an inversion in $P(a, t)$, then $\tilde{\phi}$ is the inversion in $P(\tilde{a}, t)$. If $x \in \mathbb{R}^{n-1}$, then

$$\tilde{\phi}(\tilde{x}) = \tilde{\phi}(x, 0) = (\phi(x), 0) = \widetilde{\phi(x)}.$$

The function $\tilde{\phi}$ is called the **Poincaré extension** of ϕ.

Suppose ϕ is an inversion in $S(a, r)$, $a \in \mathbb{R}^{n-1}$. Then for $x \in \mathbb{R}^n$,

$$\tilde{\phi}(x) = \tilde{a} + r^2(x - \tilde{a})^*.$$

If $[\tilde{\phi}(x)]_j$ denotes the jth coordinate function of $\tilde{\phi}(x)$, then

$$[\tilde{\phi}(x)]_n = \frac{r^2 x_n}{|x - \tilde{a}|^2}.$$

By (1.3.2) and the above

$$\frac{|\tilde{\phi}(x) - \tilde{\phi}(y)|^2}{[\tilde{\phi}(x)]_n[\tilde{\phi}(y)]_n} = \frac{|x - y|^2}{x_n y_n}.$$

As a consequence the mapping $\tilde{\phi}$ leaves

$$\frac{|x - y|^2}{x_n y_n} \tag{2.3.1}$$

invariant. If ϕ is reflection in the plane $P(a, t)$, then $\tilde{\phi}$ is a Euclidean isometry of $\mathbb{R}^n$ with $[\tilde{\phi}(x)]_n = x_n$. Thus (2.3.1) is also invariant under $\tilde{\phi}$. Furthermore, in both cases we have $\tilde{\phi}(\mathbb{H}) = \mathbb{H}$.

If ϕ is any Möbius transformation acting on $\hat{\mathbb{R}}^{n-1}$, i.e., $\phi = \phi_1 \circ \cdots \circ \phi_m$, where each ϕ_j is an inversion in $\mathbb{R}^{n-1}$, then $\tilde{\phi} = \tilde{\phi}_1 \circ \cdots \circ \tilde{\phi}_m$ is an extension of ϕ to a Möbius transformation of $\hat{\mathbb{R}}^n$ which preserves $\mathbb{H}$. By Theorem 3.2.4 of [11] this extension is unique. Also, since each $\tilde{\phi}_j$ leaves (2.3.1) invariant, so does the mapping $\tilde{\phi}$. As a consequence the Poincaré extension $\tilde{\phi}$ of any $\phi \in GM(\hat{\mathbb{R}}^{n-1})$ is an isometry of the half-space $\mathbb{H}_n$ when endowed with the Riemannian metric d given by

$$ds = \frac{|dx|}{x_n}.$$

This metric is invariant under $\tilde{\phi}$ for each $\phi \in GM(\hat{\mathbb{R}}^{n-1})$. For this metric, we have for $x, y \in \mathbb{H}_n$,

$$\sinh^2 \tfrac{1}{2} d_{\mathbb{H}}(x, y) = \frac{|x - y|^2}{4x_n y_n},$$

or

$$d_{\mathbb{H}}(x, y) = 2 \operatorname{Arcsinh} \left[\frac{|x - y|}{2\sqrt{x_n y_n}} \right].$$

We conclude this section by considering the Möbius transformation Φ that maps $\mathbb{H}$ onto $\mathbb{B}$. Set $e_n = (0, \ldots, 0, 1)$ and let Φ denote the inversion in $S(-e_n, \sqrt{2})$, that is,

$$\Phi(x) = -e_n + \frac{2(x + e_n)}{|x + e_n|^2}. \tag{2.3.2}$$

Then

$$\begin{aligned} |\Phi(x)|^2 &= 1 + \frac{4}{|x + e_n|^2} - \frac{4\langle e_n, x + e_n \rangle}{|x + e_n|^2} \\ &= 1 - \frac{4x_n}{|x + e_n|^2}, \end{aligned}$$

or

$$1 - |\Phi(x)|^2 = \frac{4x_n}{|x + e_n|^2}. \tag{2.3.3}$$

Since Φ is an inversion, Φ is a one-to-one map of $\hat{\mathbb{R}}^n$ onto $\hat{\mathbb{R}}^n$ satisfying $\Phi(\Phi(x)) = x$ for all $x \in \mathbb{R}^n$. As a consequence of (2.3.3), Φ maps $\mathbb{H}$ onto $\mathbb{B}$ and $\mathbb{B}$ onto $\mathbb{H}$. Also, since $|\Phi(x)| = 1$ when $x_n = 0$, Φ maps $\partial\mathbb{H}$ (in $\hat{\mathbb{R}}^n$) onto $\mathbb{S}$.

Since Φ is the inversion in $S(-e_n, \sqrt{2})$, by identity (1.3.2)

$$|\Phi(y) - \Phi(x)| = \frac{2|y - x|}{|y + e_n||x + e_n|},$$

and thus

$$\lim_{y \to x} \frac{|\Phi(y) - \Phi(x)|}{|y - x|} = \frac{2}{|x + e_n|^2} \tag{2.3.4}$$

$$= \frac{1 - |\Phi(x)|^2}{2x_n}. \tag{2.3.5}$$

For $\phi(x) = x^*$, we have

$$\phi'(x) = |x|^{-4}[|x|^2 I - 2Q(x)],$$

where $Q(x)$ is the $n \times n$ symmetric matrix $(x_i x_j)_{i,j=1}^n$. Since the characteristic polynomial of $2Q(x)$ is $s^{n-1}(s - 2|x|^2)$, taking $s = |x|^2$ gives

$$\det \phi'(x) = -|x|^{-2n}.$$

But $\Phi(x) = -e_n + 2\phi(x + e_n)$. Thus $\Phi'(x) = 2\phi'(x + e_n)$ and hence

$$J_\Phi(x) = \det \Phi'(x) = -\frac{2^n}{|x + e_n|^{2n}}. \tag{2.3.6}$$

2.4 Exercises

2.4.1. (a) With $\rho(x, a)$ as defined by (2.1.5), prove that

$$(1 - |x||a|)^2 \le \rho(x, a) \le (1 + |x||a|)^2.$$

(b) If $y = \varphi_a(x)$, prove that

$$(1 - |a|^2)\frac{(1 - |x|)}{(1 + |x|)} \le (1 - |y|^2) \le \frac{(1 + |x|)}{(1 - |x|)}(1 - |a|^2).$$

2.4.2. For $x \in E(r\zeta, \delta)$, show that

$$|x - \zeta| < \frac{1 + \delta}{1 - \delta}(1 - r).$$

2.4.3. Let $a \in \mathbb{B}$, $\eta \in \mathbb{S}$, and $\zeta = \varphi_a(\eta)$. Prove that

$$|a - \zeta| = \frac{(1 - |a|^2)}{|a - \eta|}.$$

Hint: Consider $|a - \zeta|^2$ and use (2.1.6).

2.4.4. Show that $|\varphi_y(x)|^2 = \dfrac{|x - y|^2}{\rho(x, y)}$.

2.4.5. If $A \in O(n)$, prove that $A \circ \varphi_a = \varphi_{Aa} \circ B$ for some $B \in O(n)$.

Exercises on the Upper Half-Space $\mathbb{H}$

2.4.6. Let Φ be the mapping defined by (2.3.2).

(a) Show that

$$\Phi(x_1, \ldots, x_n) = \left(\frac{2x_1}{|x + e_n|^2}, \ldots, \frac{2x_{n-1}}{|x + e_n|^2}, \frac{1 - |x|^2}{|x + e_n|^2}\right).$$

(b) Prove the following.

i. Φ is a one-to-one map of $\mathbb{R}^n \setminus \{e_n\}$ onto itself.

ii. $\Phi(\Phi(x)) = x$.

iii. Φ maps $\mathbb{B}$ onto $\mathbb{H}$ and $\mathbb{H}$ onto $\mathbb{B}$.

iv. Φ maps $\mathbb{S} \setminus \{e_n\}$ onto $\mathbb{R}^{n-1}$ and $\mathbb{R}^{n-1}$ onto $\mathbb{S} \setminus \{e_n\}$.

2.4.7. Show that if $n = 2$ and $z = x_1 + ix_2$, $e_2 = i$,

$$\Phi(z) = \frac{1 - i\overline{z}}{\overline{z} - i}$$

for every $z \in \mathbb{C} \setminus \{-i\}$.

2.4.8. **Möbius transformations of $\mathbb{H}$.** Let Ψ be a one-to-one map of $\mathbb{H}$ onto $\mathbb{H}$. Prove that

$$\Psi(\Phi(x)) = \Phi(A\varphi_a(x))$$

for some $a \in \mathbb{B}$ and $A \in O(n)$.

2.4.9. For y, $y^1 \in \mathbb{H}$, set

$$\Psi_{y^1}(y) = \Phi(\varphi_{\Phi(y^1)}(\Phi(y))).$$

Show that Ψ_{y^1} is a one-to-one map of $\mathbb{H}$ onto $\mathbb{H}$ with

(a) $\Psi_{y^1}(y^1) = e_n$,
(b) $\Psi_{y^1}(e_n) = y^1$, and
(c) $\Psi_{y^1}(\Psi_{y^1}(y)) = y$.

3

The Invariant Laplacian, Gradient, and Measure

In order to study harmonic function theory on the hyperbolic ball $\mathbb{B}$, we first need to determine the Laplacian Δ_h, gradient ∇^h, and measure τ on $\mathbb{B}$ that are invariant under the group $\mathcal{M}(\mathbb{B})$ of Möbius transformations of $\mathbb{B}$. Although these are well known in the setting of rank one noncompact symmetric spaces, we follow the approach of Rudin [72, Chapter 4] to determine Δ_h and ∇^h.

3.1 The Invariant Laplacian and Gradient

Definition 3.1.1 *Suppose Ω is an open subset of $\mathbb{B}$, $f \in C^2(\Omega)$, and $a \in \Omega$. We define*

$$(\Delta_h f)(a) = \Delta(f \circ \varphi_a)(0),$$

*where φ_a is the involution defined by (2.1.4) and $\Delta = \sum \frac{\partial^2}{\partial x_j^2}$ is the usual Laplacian on $\mathbb{R}^n$. The operator Δ_h is called the **invariant Laplacian** or **Laplace–Beltrami** operator on $\mathbb{B}$. Also, for $f \in C^1(\Omega)$, we define the **invariant gradient** ∇^h by*

$$(\nabla^h f)(a) = -\nabla(f \circ \varphi_a)(0),$$

where $\nabla = \left(\frac{\partial}{\partial x_1}, \ldots, \frac{\partial}{\partial x_n}\right)$ is the usual gradient.[1]

Suppose f is a C^1 (or C^2) function on $\mathbb{B}$ and $y = \psi(x)$ is a C^1 (or C^2) mapping of $\mathbb{B}$ into $\mathbb{B}$. Then if $g = f \circ \psi$, by the chain rule

$$\nabla g(x) = \nabla f(\psi(x))\psi'(x),$$

[1] The choice of the minus sign in the definition of ∇^h will assure that $\nabla^h f$ is in the same direction as ∇f. Throughout these notes we will use the notations $\widetilde{\Delta}$ and $\widetilde{\nabla}$ to denote the Möbius invariant Laplacian and gradient on the unit ball $\mathbb{B}$ of $\mathbb{C}^m$. The use of the term "invariant" is made clear in Theorem 3.1.2.

where ψ' is the $n \times n$ matrix $\left(\frac{\partial y_j}{\partial x_k}\right)$. Also

$$\Delta g(x) = \sum_{i,j=1}^{n} \frac{\partial^2 f}{\partial y_i \partial y_j} \langle \nabla y_i, \nabla y_j \rangle + \sum_{j=1}^{n} \frac{\partial f}{\partial y_j} \Delta y_j.$$

Hence if $y = \varphi_a(x)$, from Equations (2.1.8) it now follows that

$$\nabla^h f(a) = (1 - |a|^2) \nabla f(a) \tag{3.1.1}$$

and

$$\Delta_h f(a) = (1 - |a|^2)^2 \Delta f(a) + 2(n-2)(1 - |a|^2) \langle a, \nabla f(a) \rangle. \tag{3.1.2}$$

If f is a radial function, that is, $f(x) = g(|x|)$, then with $r = |x|$,

$$\nabla^h f(x) = (1 - r^2) g'(r) \frac{x}{r}, \tag{3.1.3}$$

and thus $|\nabla^h f(x)| = (1 - r^2)|g'(r)|$. Also,

$$\Delta_h f(x) = (1 - r^2) \left[(1 - r^2) g''(r) + \frac{g'(r)}{r} \{(n-1)(1 - r^2) + 2(n-2) r^2\} \right]. \tag{3.1.4}$$

The **Möbius invariant Laplacian** on the Hermitian ball $\mathbb{B}$ in $\mathbb{C}^n$ (see [72, Chapter 4]) is given by

$$\widetilde{\Delta} f(z) = 4(1 - |z|^2) \sum_{i,j=1}^{n} (\delta_{i,j} - z_i \bar{z}_j) \frac{\partial^2 f(z)}{\partial z_i \partial \bar{z}_j}.$$

In contrast to the Laplacian on real-hyperbolic space, this operator has no linear terms. The following theorem justifies the term "invariant" in reference to the operator Δ_h and the gradient ∇^h.

Theorem 3.1.2 *For $f \in C^2(\Omega)$ and $\psi \in \mathcal{M}(\mathbb{B})$,*

$$\Delta_h (f \circ \psi) = (\Delta_h f) \circ \psi \quad \textit{and} \quad |\nabla^h (f \circ \psi)| = |(\nabla^h f) \circ \psi|.$$

Proof. As in [72, Theorem 4.12], let $b \in \mathbb{B}$ and put $a = \psi(b)$. Then $\varphi_a \circ \psi \circ \varphi_b$ is a Möbius transformation of $\mathbb{B}$ fixing 0. Thus $\psi \circ \varphi_b = \varphi_a \circ A$ for some orthogonal transformation A. Hence

$$\Delta_h (f \circ \psi)(b) = \Delta (f \circ \psi \circ \varphi_b)(0) = \Delta (f \circ \varphi_a \circ A)(0).$$

But for any orthogonal transformation A, by the computations following Definition 3.1.1,

$$\Delta (g \circ A)(0) = (\Delta g)(0).$$

Therefore, $\Delta(f \circ \varphi_a \circ A)(0) = \Delta(f \circ \varphi_a)(0) = \Delta_h f(a)$, that is, $\Delta_h(f \circ \psi)(b) = (\Delta_h f)(\psi(b))$. An analogous argument proves that $|\nabla^h(f \circ \psi)| = |(\nabla^h f) \circ \psi|$. □

Remark 3.1.3 *If the hyperbolic metric (2.2.1) is expressed in standard Riemannian notation as*

$$ds^2 = \sum_{i,j} g_{i,j} dx_i dx_j,$$

where $g_{i,j} = 4\delta_{i,j}(1 - |x|^2)^{-2}$, then the Laplace–Beltrami operator L on $\mathbb{B}$ is defined by

$$L(f) = \frac{1}{\sqrt{g}} \sum_{i,j} \partial_i (g^{i,j} \sqrt{g} \partial_j f), \tag{3.1.5}$$

where $g = \det(g_{i,j})$, and $(g^{i,j})$ is the inverse matrix of $(g_{i,j})$. A brief computation shows that except for a factor of $\frac{1}{4}$, this agrees with the operator Δ_h given by (3.1.2). Likewise, the gradient of a function f is the vector field grad f *defined by*

$$\operatorname{grad} f = \sum g^{i,j} \frac{\partial f}{\partial x_i} \frac{\partial}{\partial x_j}. \tag{3.1.6}$$

In particular

$$\operatorname{grad} f = \frac{1}{4}(1 - |x|^2)^2 \sum_{j=1}^{n} \frac{\partial f}{\partial x_j} \frac{\partial}{\partial x_j},$$

and

$$\langle \operatorname{grad} f, \operatorname{grad} g \rangle = (\operatorname{grad} f) g = (\operatorname{grad} g) f = \frac{1}{4}(1 - |x|^2)^2 \langle \nabla f, \nabla g \rangle.$$

Hence

$$|\operatorname{grad} f| = \sqrt{\langle \operatorname{grad} f, \operatorname{grad} f \rangle} = \frac{1}{2}(1 - |x|^2)|\nabla f|.$$

Thus, except for the factor of $\frac{1}{2}$, $|\operatorname{grad} f|$ *agrees with $|\nabla^h f|$ as given by (3.1.1).*

3.2 The Fundamental Solution of Δ_h

Suppose g is a radial solution of $\Delta_h g = 0$. If we let $v(r) = g'(r)$, then by (3.1.4) the function v must satisfy

$$(1 - r^2)v'(r) + \{(n-1)(1 - r^2) + 2(n-2)r^2\} \frac{v(r)}{r} = 0$$

or

$$\frac{v'(r)}{v(r)} = -(n-1)\frac{1}{r} - (n-2)\frac{2r}{(1-r^2)}.$$

Solving this differential equation for $v(r)$ gives

$$v(r) = c\frac{(1-r^2)^{n-2}}{r^{n-1}}$$

for some constant c. Thus the **fundamental solution** g_h for Δ_h is given by

$$g_h(r) = \frac{1}{n}\int_r^1 \frac{(1-s^2)^{n-2}}{s^{n-1}}ds. \tag{3.2.1}$$

The choice of the constant $\frac{1}{n}$ will become apparent in Theorem 4.1.1. When $n = 2$, this gives the usual solution

$$g_h(r) = \frac{1}{2}\log\frac{1}{r}.$$

Using the inequality

$$\frac{r}{1+r} < \log(1+r) < r, \quad r > -1,\ r \neq 0,$$

we have that when $n = 2$, the fundamental solution $g(|x|) = -\frac{1}{2}\log|x|$ satisfies

$$\frac{1}{4}(1-|x|^2) \le g_h(|x|) \le \frac{1}{2}\frac{(1-|x|^2)}{|x|}. \tag{3.2.2}$$

When $n > 2$, we estimate the integral (3.2.1) above and below as follows. For an upper estimate we have

$$\begin{aligned}\int_r^1 \frac{(1-s^2)^{n-2}}{s^{n-1}}ds &\le (1-r^2)^{n-2}\int_r^1 \frac{ds}{s^{n-1}}\\ &= \frac{1}{(n-2)}(1-r^2)^{n-2}\left[\frac{1}{r^{n-2}} - 1\right].\end{aligned}$$

But

$$\frac{1}{r^{n-2}} - 1 = \left[\frac{1-r^{n-2}}{r^{n-2}}\right] \le (n-2)\frac{(1-r)}{r^{n-2}} \le (n-2)\frac{(1-r^2)}{r^{n-2}}.$$

Therefore, $g_h(r) \le \frac{1}{n}(1-r^2)^{n-1}r^{2-n}$. For a lower estimate, we have

$$\begin{aligned}\frac{1}{n}\int_r^1 \frac{(1-s^2)^{n-2}}{s^{n-1}}ds &= \frac{1}{n}\int_r^1 \frac{(1-s^2)^{n-2}s}{s^n}ds\\ &\ge \frac{1}{n}\int_r^1 (1-s^2)^{n-2}s\,ds = \frac{1}{2n(n-1)}(1-r^2)^{n-1}.\end{aligned}$$

Therefore, $g_h(r) \geq \frac{1}{2n(n-1)}(1-r^2)^{n-1}$. Thus the fundamental solution $g_h(|x|)$ satisfies

$$\frac{1}{2n(n-1)}(1-|x|^2)^{n-1} \leq g_h(|x|) \leq \frac{1}{n}\frac{(1-|x|^2)^{n-1}}{|x|^{n-2}} \tag{3.2.3}$$

for all $x \in \mathbb{B}$, $x \neq 0$.

Definition 3.2.1 *For $x, y \in \mathbb{B}$, $x \neq y$, the **Green's function** $G_h(x,y)$ for Δ_h is defined by*

$$G_h(x,y) = g_h(|\varphi_y(x)|) = \frac{1}{n}\int_{|\varphi_y(x)|}^{1} \frac{(1-s^2)^{n-2}}{s^{n-1}}\, ds.$$

Since $|\varphi_y(x)| = |\varphi_x(y)|$, we have that $G_h(x,y) = G_h(y,x)$ for all x, y with $x \neq y$. Also, if $y \in \mathbb{B}$ is fixed, the function $x \mapsto G_h(x,y)$ satisfies $\Delta_h G_h(x,y) = 0$ on $\mathbb{B} \setminus \{y\}$. Furthermore, by (3.2.3), for $n > 2$,

$$c_n(1-|\varphi_y(x)|^2)^{n-1} \leq G_h(x,y) \leq \frac{1}{n}\frac{(1-|\varphi_y(x)|^2)^{n-1}}{|\varphi_y(x)|^{n-2}} \tag{3.2.4}$$

for all $x, y \in \mathbb{B}$, $x \neq y$ with $c_n = \frac{1}{2n(n-1)}$. Also, by (2.1.6) and (2.1.7), for $n > 2$ we have

$$G_h(x,y) \leq \frac{1}{n}\frac{(1-|x|^2)^{n-1}(1-|y|^2)^{n-1}}{|x-y|^{n-2}\{|x-y|^2 + (1-|x|^2)(1-|y|^2)\}^{n/2}}, \tag{3.2.5}$$

and

$$G_h(x,y) \geq c_n \frac{(1-|x|^2)^{n-1}(1-|y|^2)^{n-1}}{(1+|x||y|)^{2(n-1)}}. \tag{3.2.6}$$

In comparison, the **Euclidean Green's function**, G^e, for the Laplacian Δ on $\mathbb{B}$ is given for $n \geq 3$ by

$$G^e(x,y) = c_n \begin{cases} |x-y|^{2-n} - (|y||x-y^*|)^{2-n}, & y \in \mathbb{B}\setminus\{0\}, \\ |x|^{2-n} - 1, & y = 0, \end{cases} \tag{3.2.7}$$

for a suitable constant c_n. When $n = 2$, we have

$$G^e(x,y) = G_h(x,y) = \frac{1}{2}\log\left(\frac{|y||x-y^*|}{|x-y|}\right). \tag{3.2.8}$$

3.3 The Invariant Measure on $\mathbb{B}$

Our next step is to determine the Möbius invariant measure τ on $\mathbb{B}$. First, however, we introduce some notation and formulas concerning integration on $\mathbb{B}$. We denote by ν a Lebesgue measure in $\mathbb{R}^n$ normalized so that $\nu(\mathbb{B}) = 1$. Also, we denote by σ or σ_n surface measure on $\mathbb{S}$ again normalized such that $\sigma(\mathbb{S}) = 1$. Then by integration in polar coordinates we have

$$\int_{\mathbb{B}} f(x)\, d\nu(x) = n \int_0^1 r^{n-1} \int_{\mathbb{S}} f(r\zeta)\, d\sigma(\zeta) dr. \tag{3.3.1}$$

The measure σ is invariant under $O(n)$, that is,

$$\int_{\mathbb{S}} f(A\zeta) d\sigma(\zeta) = \int_{\mathbb{S}} f(\zeta) d\sigma(\zeta) \tag{3.3.2}$$

for all $A \in O(n)$ and $f \in L^1(\mathbb{S})$. Furthermore, if K is any compact subgroup of $O(n)$,

$$\int_{\mathbb{S}} f(\zeta) d\sigma(\zeta) = \int_{\mathbb{S}} \int_K f(k\zeta)\, dk d\sigma(\zeta), \tag{3.3.3}$$

where dk denotes **Haar measure**[2] on K. In particular, with $K = O(n)$ one obtains

$$\int_{\mathbb{S}} f(\zeta) d\sigma(\zeta) = \int_{O(n)} f(A\zeta) dA, \tag{3.3.4}$$

where dA is the Haar measure on $O(n)$.

To determine the invariant measure τ on $\mathbb{B}$ we assume that $d\tau(x) = \rho(x) d\nu(x)$, where ρ is a radial function on $\mathbb{B}$. Then for $f \in L^1(\mathbb{B}, \tau)$ and the mapping $\varphi_a \in \mathcal{M}(\mathbb{B})$ we have

$$\int_{\mathbb{B}} (f \circ \varphi_a) d\tau = \int_{\mathbb{B}} f(\varphi_a(x)) \rho(x) d\nu(x),$$

which by the change of variables formula for $\mathbb{R}^n$

$$= \int_{\mathbb{B}} f(x) \rho(\varphi_a(x)) |J_{\varphi_a}(x)| d\nu(x),$$

where J_{φ_a} is the Jacobian of the mapping φ_a. Thus in order that

$$\int (f \circ \varphi_a) d\tau = \int f d\tau$$

for all $f \in L^1(\mathbb{B}, \tau)$ and φ_a, we must have $\rho(\varphi_a(x)) |J_{\varphi_a}(x)| = \rho(x)$. In particular with $x = 0$, by (2.1.8), $\rho(a) = (1 - |a|^2)^{-n} \rho(0)$. Hence we define the measure τ on $\mathbb{B}$ by

[2] A Borel measure k on a compact group K is said to be a **Haar measure** if $k(K) = 1$ and

$$\int_K f(kk_1) dk = \int_K f(k) dk = \int_K f(k_1 k) dk$$

for each $k_1 \in K$ and $f \in C(K)$.

$$d\tau(x) = \frac{d\nu(x)}{(1-|x|^2)^n}. \tag{3.3.5}$$

In the following theorem we prove that τ is the **Möbius invariant measure** on $\mathbb{B}$.

Theorem 3.3.1 **(a)** *If $\psi \in \mathcal{M}(\mathbb{B})$, then the Jacobian J_ψ of ψ satisfies*

$$|J_\psi(x)| = \frac{(1-|\psi(x)|^2)^n}{(1-|x|^2)^n}$$

for all $x \in \mathbb{B}$.

(b) *The measure τ defined by (3.3.5) satisfies*

$$\int_{\mathbb{B}} f\, d\tau = \int_{\mathbb{B}} (f \circ \psi)\, d\tau$$

for every $f \in L^1(\mathbb{B}, \tau)$ and $\psi \in \mathcal{M}(\mathbb{B})$.

Proof. (a) For $\psi \in \mathcal{M}(\mathbb{B})$ and $a \in \mathbb{B}$, let $b = \psi(a)$. Then $\varphi_b \circ \psi \circ \varphi_a \in \mathcal{M}(\mathbb{B})$ and fixes 0. Thus $\psi(x) = \varphi_b \circ A \circ \varphi_a(x)$ for some $A \in O(n)$. But then $\psi'(x) = \varphi_b'(A\varphi_a(x))A\varphi_a'(x)$, or

$$\psi'(a) = \varphi_b'(0)A\varphi_a'(a).$$

Hence by (2.1.8)

$$\psi'(a) = \frac{(1-|b|^2)}{(1-|a|^2)}A.$$

Since A is orthogonal, $|\det A| = 1$. Therefore

$$|J_\psi(a)| = |\det \psi'(a)| = \frac{(1-|\psi(a)|^2)^n}{(1-|a|^2)^n}.$$

For (b), if $f \in L^1(\mathbb{B}, \tau)$ and $\psi \in \mathcal{M}(\mathbb{B})$, then

$$\int_{\mathbb{B}} f\, d\tau = \int_{\mathbb{B}} f(w)(1-|w|^2)^{-n} d\nu(w),$$

which by the change of variables formula

$$\begin{aligned} &= \int_{\mathbb{B}} f(\psi(x))(1-|\psi(x)|^2)^{-n}|J_\psi(x)|d\nu(x) \\ &= \int_{\mathbb{B}} f(\psi(x))(1-|x|^2)^{-n}d\nu(x) = \int_{\mathbb{B}} (f\circ\psi)\, d\tau. \end{aligned}$$

□

For future reference we estimate $\tau(E(a,r))$, $0 < r < 1$. Since $E(a,r) = \varphi_a(B_r)$,

$$\tau(E(a,r)) = \tau(B_r) = n\int_0^r \frac{\rho^{n-1}}{(1-\rho^2)^n}d\rho. \tag{3.3.6}$$

For $n \geq 2$, set

$$c_n(r) = \frac{r^n}{(1-r^2)^{n-1}}. \tag{3.3.7}$$

Then $c_2(r) = \tau(E(a,r))$, and for $n > 2$, by L'Hospital's rule,

$$\lim_{r\to 1^-} \frac{\tau(B_r)}{c_n(r)} = \lim_{r\to 1^-} \frac{n}{n(1-r^2)+2(n-1)r^2} = \frac{n}{2(n-1)}.$$

Likewise,

$$\lim_{r\to 0^+} \frac{\tau(B_r)}{c_n(r)} = 1.$$

Thus for all $n \geq 2$,

$$\tau(B_r) = \tau(E(a,r)) \approx^3 \frac{r^n}{(1-r^2)^{n-1}}, \tag{3.3.8}$$

with equality when $n = 2$.

3.4 The Invariant Convolution on $\mathbb{B}$

For $0 < p < \infty$, we denote by $L^p(\mathbb{B},\tau)$ the space of measurable functions f on $\mathbb{B}$ for which

$$\|f\|_p^p = \int_{\mathbb{B}} |f(x)|^p \, d\tau(x) < \infty.$$

Also, $L^p_{loc}(\mathbb{B})$ will denote the space of measurable functions f on $\mathbb{B}$ that are **locally p-integrable**, that is,

$$\int_K |f(x)|^p \, d\tau(x) < \infty$$

for every compact subset K of $\mathbb{B}$.[4]

Definition 3.4.1 *For measurable functions f, g on $\mathbb{B}$, the* ***invariant convolution*** *$f * g$ of f and g is defined by*

[3] $A(r) \approx B(r)$ means that there exist positive constants c_1 and c_2 such that $c_1 A(r) \leq B(r) \leq c_2 A(r)$ for all r.

[4] It should be obvious that in the definition of local integrability the measure τ may be replaced by the Lebesgue measure ν.

$$(f * g)(y) = \int_{\mathbb{B}} f(x)g(\varphi_y(x))\, d\tau(x), \qquad y \in \mathbb{B},$$

provided this integral exists.

By the invariance of τ we have $(f * g)(y) = (g * f)(y)$. Although the convolution as defined is not the usual definition for convolution of functions on a topological group, the following analogue of the standard convolution inequalities is still valid.

Theorem 3.4.2 *Let $p \in [1, +\infty)$ and let q be defined by $\frac{1}{p} + \frac{1}{q} = 1$. If $f \in L^p(\mathbb{B}, \tau)$, then*

$$\|f * g\|_p \le \|f\|_p \|g\|_1$$

for all radial functions $g \in L^1(\mathbb{B}, \tau)$, and

$$\|f * g\|_\infty \le \|f\|_p \|g\|_q$$

for all radial functions $g \in L^q(\mathbb{B}, \tau)$.

Proof. Let $g \in L^1(\mathbb{B}, \tau)$ be a radial function, and $h \in L^q(\mathbb{B}, \tau)$. Without loss of generality we may assume that f, g and h are non-negative. Thus by Definition 3.4.1 and Tonelli's theorem

$$\begin{aligned}\int_{\mathbb{B}} f(x)(f * g)(x) d\tau(x) &= \int_{\mathbb{B}} h(x) \int_{\mathbb{B}} f(y)(g \circ \varphi_x)(y) d\tau(y) d\tau(x) \\ &= \int_{\mathbb{B}} f(y) \int_{\mathbb{B}} h(x)(g(\varphi_x(y)) d\tau(x) d\tau(y).\end{aligned}$$

Since $\rho(x, y) = \rho(y, x)$, by (2.1.7), we have that $|\varphi_x(y)| = |\varphi_y(x)|$. Thus since g is radial,

$$\int_{\mathbb{B}} h(x) g(\varphi_x(y)) d\tau(x) = \int_{\mathbb{B}} h(x) g(\varphi_y(x)) d\tau(x),$$

which by the invariance of τ

$$= \int_{\mathbb{B}} g(x) h(\varphi_y(x)) d\tau(x).$$

Therefore

$$\|(f * g)h\|_1 \le \int_{\mathbb{B}} f(y) \left\{ \int_{\mathbb{B}} g^{\frac{1}{p}}(x) g^{\frac{1}{q}}(x)(h \circ \varphi_y)(x) d\tau(x) \right\} d\tau(y),$$

which by Hölder's inequality

$$\le \|g\|_1^{\frac{1}{p}} \int_{\mathbb{B}} f(y) \left\{ \int_{\mathbb{B}} g(x)(h \circ \varphi_y)^q(x) d\tau(x) \right\}^{\frac{1}{q}} d\tau(y).$$

One more application of Hölder's inequality now gives

$$\|(f * g)h\|_1 \le \|g\|_1^{\frac{1}{p}} \|f\|_p \left\{ \int_{\mathbb{B}} \int_{\mathbb{B}} g(x)(h \circ \varphi_y)^q(x) d\tau(x) d\tau(y) \right\}^{\frac{1}{q}}.$$

Also, since τ is invariant and g is radial,

$$\begin{aligned}
\int_{\mathbb{B}} \int_{\mathbb{B}} g(x)(h \circ \varphi_y)^q(x) d\tau(x) d\tau(y) &= \int_{\mathbb{B}} \int_{\mathbb{B}} (g \circ \varphi_y)(x) h^q(x) d\tau(x) d\tau(y) \\
&= \int_{\mathbb{B}} \int_{\mathbb{B}} (g \circ \varphi_x)(y) h^q(x) d\tau(y) d\tau(x) \\
&= \int_{\mathbb{B}} g(y) d\tau(y) \int_{\mathbb{B}} h^q(x) d\tau(x).
\end{aligned}$$

Therefore,

$$\|(f * g)h\|_1 \le \|g\|_1^{\frac{1}{p}} \|f\|_p \|g\|_1^{\frac{1}{q}} \|h\|_q.$$

Taking the supremum over all h with $\|h\|_q \le 1$ gives

$$\|f * g\|_p \le \|f\|_p \|g\|_1.$$

The inequality $\|f * g\|_\infty \le \|f\|_p \|g\|_q$ is an easy consequence of Hölder's inequality and the invariance of the measure τ. □

As a consequence of Theorem 3.4.2, if $g \in L^1_{loc}(\mathbb{B})$ is radial and $f \in L^p_{loc}(\mathbb{B})$ then $f * g$ is defined a.e. on $\mathbb{B}$. There is one additional property of the above convolution that will be needed in the proof of Theorem 4.5.4. The analogue of the theorem for functions in the unit ball in $\mathbb{C}^n$ is due to Ullrich [94].

Theorem 3.4.3 *If $f, \chi, h \in L^1(\mathbb{B}, \tau)$ and χ is radial, then $(f * \chi) * h = f * (\chi * h)$.*

Proof. Suppose $a, y \in \mathbb{B}$. Since $\varphi_{\varphi_a(y)} \circ \varphi_a \circ \varphi_y(0) = 0$, by Theorem 2.1.1

$$\varphi_{\varphi_a(y)} \circ \varphi_a \circ \varphi_y = A$$

for some $A \in O(n)$. Therefore

$$\varphi_{\varphi_a(y)} = A \circ \varphi_y \circ \varphi_a.$$

Consequently $|\varphi_{\varphi_a(y)}(x)| = |\varphi_y(\varphi_a(x))|$. By the invariance of τ and the fact that χ is radial,

$$\begin{aligned}\int_{\mathbb{B}} \chi(x)h(\varphi_{\varphi_a(y)}(x))d\tau(x) &= \int_{\mathbb{B}} \chi(\varphi_{\varphi_a(y)}(x))h(x)d\tau(x) \\ &= \int_{\mathbb{B}} \chi(\varphi_y(\varphi_a(x)))h(x)d\tau(x) \\ &= \int_{\mathbb{B}} \chi(\varphi_y(x))h(\varphi_a(x))d\tau(x).\end{aligned}$$

Therefore,

$$\begin{aligned}((f * \chi) * h)(a) &= \int_{\mathbb{B}}\int_{\mathbb{B}} f(x)\chi(\varphi_y(x))h(\varphi_a(y))d\tau(x)d\tau(y) \\ &= \int_{\mathbb{B}}\int_{\mathbb{B}} f(x)\chi(y)h(\varphi_{\varphi_a(x)}(y))d\tau(y)d\tau(x) \\ &= \int_{\mathbb{B}}\int_{\mathbb{B}} f(x)(\chi * h)(\varphi_a(x))d\tau(x) \\ &= (f * (\chi * h))(a).\end{aligned}$$

□

3.5 Exercises

3.5.1. For a function u defined on $\mathbb{B}$ and $0 < r < 1$, set $u_r(x) = u(rx)$.

(a) If u is Euclidean harmonic on $\mathbb{B}$, that is, $\Delta u = 0$, show that u_r is Euclidean harmonic on $\mathbb{B}$ for all $n \geq 2$.

(b) If $n \geq 3$ and u is $\mathcal{H}$-harmonic on $\mathbb{B}$, show by example that u_r need not be $\mathcal{H}$-harmonic.

3.5.2. Show that $\Delta g(Ax) = \Delta g(x)$ for all $A \in O(n)$.

3.5.3. (a) Suppose $f, g \in C^2(\mathbb{B})$. Prove that

$$\Delta_h(fg) = f\Delta_h g + g\Delta_h f + 2\langle\nabla^h f, \nabla^h g\rangle.$$

(b) If $f \in C^2(\mathbb{R})$ and $g \in C^2(\mathbb{B})$, prove that

$$\Delta_h f(g(x)) = f''(g(x))|\nabla^h g(x)|^2 + f'(g(x))\Delta_h g(x).$$

3.5.4. (a) Prove that for all r, $0 < r < 1$,

$$\frac{1}{2^n}\frac{r^n}{(1-r^2)^{n-1}} \leq \tau(B_r) \leq \frac{n}{2(n-1)}\frac{r^n}{(1-r^2)^{n-1}}.$$

(b) With $\rho = \tanh r$, prove that

$$\tau(B_r) = n \int_0^{\tanh^{-1}\rho} \sinh^{n-1} t \cosh^{n-1} t dt.$$

3.5.5. Let $f \in C^2(\mathbb{B})$ and let $Z_f = \{x \in \mathbb{B} : f(x) = 0\}$. Show that for $p > 1$

$$\Delta_h|f|^p = p(p-1)|f|^{p-2}|\nabla^h f|^2 + p|f|^{p-2} f \Delta_h f \qquad (3.5.1)$$

on $\mathbb{B} \setminus Z_f$.

3.5.6. For a C^1 function f, the **normal derivative operator** N is defined by $Nf(x) = \langle x, \nabla f(x)\rangle$. If $r = |x|$, show that

$$\Delta_h = \frac{(1-r^2)}{r^2}\left[(1-r^2)N^2 + (n-2)(1+r^2)N + (1-r^2)\Delta_\sigma\right],$$

where Δ_σ is the tangential part of the Euclidean Laplacian given in Cartesian coordinates by

$$\Delta_\sigma = \sum_{i<j} L_{i,j}^2, \quad \text{where} \quad L_{i,j} = x_i \frac{\partial}{\partial x_j} - x_j \frac{\partial}{\partial x_i}.$$

3.5.7. As in (3.2.7) and (3.2.8), let G^e denote the Euclidean Green's function on $\mathbb{B}$.

(a) Prove that

$$G^e(x,y) \le K \begin{cases} -\log|x-y|, & n = 2, \\ |x-y|^{2-n}, & n \ge 3. \end{cases}$$

(b) Prove that

$$G^e(x,y) \le K G^e(0,y)|x-y|^{1-n}. \qquad (3.5.2)$$

(c) For fixed $y_o \in \mathbb{B}$, prove that

$$G^e(x,y_o) \ge c(1-|x|) \quad \text{for all } x \in \mathbb{B},$$

and

$$G^e(x,y_o) \le c_\delta(1-|x|) \quad \text{for all } x \in \mathbb{B} \setminus B(y_o,\delta),$$

where $\delta < (1-|y_o|)$.

3.5.8. Fix δ, $0 < \delta < 1$. Prove that

$$G_h(x,y) \le \begin{cases} c_\delta G_h(0,y)\dfrac{(1-|x|^2)^{n-1}}{\rho(x,y)^{n-1}}, & \text{for all } x \in \mathbb{B} \setminus E(y,\delta), \\ c_n G_h(0,y)|x-y|^{1-n}, & \text{for all } x \in \mathbb{B}. \end{cases}$$

3.5.9. (a) Show that $|\nabla^h g(x)| = \dfrac{1}{n}\dfrac{(1-|x|^2)^{n-1}}{|x|^{n-1}}$.

(b) Using (a) compute $|\nabla_x^h G_h(x,y)|$.

3.5.10. Consider $\varphi_x(a)$ as defined by (2.1.6), that is,

$$\varphi_x(a) = \frac{x|x-a|^2 + (1-|a|^2)(x-a)}{\rho(x,a)}.$$

(a) Let y_j be the jth coordinate of $y = \varphi_x(a)$. Show that

$$\frac{\partial y_j}{\partial x_i}\Big|_{x=0} = \delta_{i,j}(|a|^2+1) - 2a_i a_j,$$

where $\delta_{i,j} = 1$ if $i = j$, 0 otherwise.

(b) If f is a C^1 function, prove that

$$\nabla_x f(\varphi_x(a))|_{x=0} = -2\langle a, \nabla f(-a)\rangle a + (1+|a|^2)\nabla f(-a). \quad (3.5.3)$$

Exercises on the Upper Half-Space $\mathbb{H}$

3.5.11. **The invariant Laplacian and gradient on $\mathbb{H}$.** Using formulas (3.1.5) and (3.1.6), show that for the upper half-space $\mathbb{H}$ with the metric $ds = y_n^{-1}|dy|$,

$$L_{\mathbb{H}} f(y) = y_n^2 \Delta f(y) - (n-2)y_n \frac{\partial f}{\partial y_n}$$

and

$$(\operatorname{grad} f)g = y_n^2\langle \nabla f, \nabla g\rangle = \langle y_n \nabla f, y_n \nabla g\rangle,$$

or $\nabla^h f(y) = y_n \nabla f(y)$.

3.5.12. Let U be a real-valued function on $\mathbb{H}$, and let $\Phi : \mathbb{B} \to \mathbb{H}$ be given by (2.3.2). Let $V(x)$ be defined on $\mathbb{B}$ by $V(x) = U(\Phi(x))$.

(a) Prove that $|\nabla^h V(x)|^2 = 4y_n^2|\nabla U(y)|^2$.

(b) Prove that $\Delta_h V(x) = 4L_{\mathbb{H}} U(y)$.

3.5.13. **The Green's function on $\mathbb{H}$.** For $y = (y', y_n) \in \mathbb{H}$, set

$$g_{\mathbb{H}}(y) = g_h(\Phi(y)),$$

where $\Phi : \mathbb{H} \to \mathbb{B}$ is given by (2.3.2). Show that

(a) $L_{\mathbb{H}} g_{\mathbb{H}}(y) = 0$ for all $y \in \mathbb{H} \setminus \{e_n\}$ with $g_{\mathbb{H}}(e_n) = +\infty$.

(b) $c_n 4^{n-1} \dfrac{y_n^{n-1}}{|y+e_n|^{2(n-1)}} \le g_{\mathbb{H}}(y) \le \dfrac{4^{n-1}}{n}\dfrac{y_n^{n-1}}{|y+e_n|^n|y-e_n|^{n-2}}$.

3.5.14. As in the previous exercise, for $y_o, y \in \mathbb{H}$, set $G_{\mathbb{H}}(y, y_o) = G_h(\Phi(y), \Phi(y_o))$. Then $G_{\mathbb{H}}$ is the Green's function for $L_{\mathbb{H}}$ on $\mathbb{H}$ with singularity at y_o. Using inequalities (3.2.5) and (3.2.6) find upper and lower estimates for $G_{\mathbb{H}}(y, y_o)$.

3.5.15. **The invariant measure on $\mathbb{H}$.** Let $f \in C_c(\mathbb{H})$, and let $\Phi : \mathbb{B} \to \mathbb{H}$ be the map given by (2.3.2).

(a) Using the change of variable formula and (2.3.6), prove that

$$\int_{\mathbb{H}} f(y)dy = 2^n \int_{\mathbb{B}} \frac{f(\Phi(x))}{|x + e_n|^{2n}} dx.$$

(b) Using (a), show that

$$\int_{\mathbb{H}} f(y)dy = 2^n \int_{\mathbb{B}} \frac{f(\Phi(x))}{(1 - |x|^2)^n} dx.$$

(c) Set $d\tau_{\mathbb{H}}(y) = y_n^{-n}dy$. If Ψ is a one-to-one map of $\mathbb{H}$ onto $\mathbb{H}$ prove that

$$\int_{\mathbb{H}} f(\Psi(y))d\tau_{\mathbb{H}}(\Psi(y)) = \int_{\mathbb{H}} f(y)d\tau_{\mathbb{H}}(y).$$

3.5.16. For any positive measure ν on $\mathbb{S}$, define $\nu^* = \nu \circ \Phi$ on the Borel subsets of $\mathbb{R}^{n-1}$ by

$$\nu^*(B) = (\nu \circ \Phi)(B) = \nu(\Phi(B)).$$

Prove that ν^* is a measure on $\mathbb{R}^{n-1}$ and

$$\int_{\mathbb{S}\setminus\{e_n\}} f d\nu = \int_{\mathbb{R}^{n-1}} (f \circ \Phi)d(\nu \circ \Phi)$$

for every positive Borel measurable function f on $\mathbb{S} \setminus \{e_n\}$.

3.5.17. **Change of variables formulas for $\mathbb{S}$.** [10, p. 148] Use the previous exercise with $\nu = \sigma$ to

(a) prove that

$$\int_{\mathbb{S}} f d\sigma = c^n 2^{n-2} \int_{\mathbb{R}^{n-1}} \frac{f(\Phi(t))}{(1 + |t|^2)^{n-1}} dt$$

for every positive Borel measurable function on $\mathbb{S}$.

(b) Prove that

$$\int_{\mathbb{R}^{n-1}} f(t)dt = \frac{2}{c_n} \int_{\mathbb{S}} \frac{f(\Phi(\zeta))}{(1 + \zeta_n)^{n-1}} d\sigma(\zeta)$$

for every positive Borel measurable function on $\mathbb{R}^{n-1}$.

4

$\mathcal{H}$-Harmonic and $\mathcal{H}$-Subharmonic Functions

In this chapter we consider the class of functions on $\mathbb{B}$ that are harmonic or subharmonic with respect to the operator Δ_h. We begin with the following definition.

Definition 4.0.1 *Let Ω be an open subset of $\mathbb{B}$. A real (or complex)-valued function $f \in C^2(\Omega)$ is said to be $\mathcal{H}$-**harmonic** (or invariant harmonic) on Ω if $\Delta_h f(x) = 0$ for all $x \in \Omega$. Also, a real-valued function $f \in C^2(\Omega)$ is said to be $\mathcal{H}$-**subharmonic** (or invariant subharmonic) on Ω if $\Delta_h f(x) \geq 0$ for all $x \in \Omega$.*[1]

Clearly if f is $\mathcal{H}$-harmonic or $\mathcal{H}$-subharmonic on $\mathbb{B}$, then by the $\mathcal{M}$-invariance of Δ_h so is $f \circ \psi$ for all $\psi \in \mathcal{M}(\mathbb{B})$. In Section 4.1 we derive the invariant mean-value property for C^2 $\mathcal{H}$-subharmonic functions, and in Section 4.3 we extend the definition of $\mathcal{H}$-subharmonic functions to the class of upper semicontinuous functions. Sections 4.4, 4.5, and 4.6 contain basic properties of $\mathcal{H}$-subharmonic functions. We conclude the chapter with a discussion of quasi-nearly $\mathcal{H}$-subharmonic functions, a concept that will play a significant role in subsequent chapters.

4.1 The Invariant Mean-Value Property

Recall that a function f on $\mathbb{B}$ is **radial** if $f(x) = g(|x|)$ for some function g on $[0, 1)$. This is equivalent to $f(Ax) = f(x)$ for all $A \in O(n)$. For a continuous function f on $\mathbb{B}$ we define the **radialization** $f^{\#}$ of f by

[1] In the text by W. Rudin ([72]) functions that are harmonic or subharmonic with respect to the Möbius invariant Laplacian are referred to as $\mathcal{M}$-harmonic and $\mathcal{M}$-subharmonic. In these notes I use the terms $\mathcal{H}$-harmonic and $\mathcal{H}$-subharmonic to denote functions that are harmonic or subharmonic with respect to the hyperbolic Laplacian.

$$f^\sharp(x) = \int_{O(n)} f(Ax)\, dA = \int_{\mathbb{S}} f(|x|\zeta) d\sigma(\zeta), \quad x \in \mathbb{B}, \tag{4.1.1}$$

where dA is the Haar measure on $O(n)$. Clearly $f^\sharp$ is a radial function on $\mathbb{B}$, and if $f \in C^2(\mathbb{B})$,

$$\Delta_h f^\sharp(x) = \int_{O(n)} (\Delta_h f)(Ax) dA.$$

Prior to proving the mean-value property for $\mathcal{H}$-harmonic and $\mathcal{H}$-subharmonic functions we prove the following analogue of Lemma 2.5 of [62].

Theorem 4.1.1 *If $f \in C^2(\mathbb{B})$, then*

(a) $\displaystyle \frac{d}{dr}\int_{\mathbb{S}} f(r\zeta) d\sigma(\zeta) = \tfrac{1}{n} r^{1-n}(1-r^2)^{n-2} \int_{B_r} \Delta_h f(x) d\tau(x)$, *and*

(b) $\displaystyle f(0) = \int_{\mathbb{S}} f(r\zeta) d\sigma(\zeta) - \int_{B_r} g(|x|, r) \Delta_h f(x)\, d\tau(x)$, *where*

$$g(|x|, r) = \frac{1}{n}\int_{|x|}^{r} \frac{(1-s^2)^{n-2}}{s^{n-1}}\, ds. \tag{4.1.2}$$

Proof. Since $(\Delta_h f)^\sharp = \Delta_h(f^\sharp)$, to prove (a) it suffices to assume that f is a radial function on $\mathbb{B}$. As in [62], suppose $f(x) = u(|x|^2)$, where u is a C^2 function on $[0, 1)$. Then with $r = |x|$,

$$\Delta f(x) = 4r^2 u''(r^2) + 2nu'(r^2) \quad \text{and} \quad \langle x, \nabla f(x)\rangle = 2r^2 u'(r^2).$$

Thus by (3.1.2)

$$\Delta_h f(x) = 4r^2(1-r^2)^2 u''(r^2) + 2n(1-r^2)^2 u'(r^2) + 4(n-2)r^2(1-r^2)u'(r^2). \tag{4.1.3}$$

If we let

$$v(t) = t^{\frac{n}{2}}(1-t)^{2-n} u'(t),$$

then $\Delta_h f(x) = 4r^{2-n}(1-r^2)^n v'(r^2)$. Therefore

$$\begin{aligned}
\int_{B_\rho} \Delta_h f\, d\tau &= n\int_0^\rho r^{n-1}(1-r^2)^{-n} \Delta_h f(r)\, dr \\
&= 4n\int_0^\rho r v'(r^2)\, dr \\
&= 2n\rho^n(1-\rho^2)^{2-n} u'(\rho^2).
\end{aligned}$$

On the other hand,

$$\frac{d}{d\rho}\int_{\mathbb{S}} f(\rho\zeta)\, d\sigma(\zeta) = 2\rho u'(\rho^2).$$

Therefore

$$\frac{d}{d\rho}\int_{\mathbb{S}} f(\rho\zeta)\,d\sigma(\zeta) = \frac{1}{n}\rho^{1-n}(1-\rho^2)^{n-2}\int_{B_\rho} \Delta_h f(x)\,d\tau(x),$$

which proves (a). Integrating the above from 0 to r gives

$$\int_{\mathbb{S}} f(r\zeta)d\sigma(\zeta) - f(0) = \int_0^r \frac{1}{n}\rho^{1-n}(1-\rho^2)^{n-2}\int_{B_\rho} \Delta_h f(x)\,d\tau(x)\,d\rho,$$

which upon changing the order of integration

$$= \int_{B_r} g(|x|, r)\Delta_h f(x)\,d\tau(x),$$

where $g(|x|, r)$ is defined by (4.1.2). □

Remark 4.1.2 *Although we assumed in Theorem 4.1.1 that $f \in C^2(\mathbb{B})$, the conclusions are still valid if $f \in C^2(\Omega)$, where Ω is an open subset of $\mathbb{B}$ with $0 \in \Omega$. In this case the conclusions hold for all r such that $B(0, r) \subset \Omega$.*

As a corollary of Theorem 4.1.1 we obtain the following mean-value property for $\mathcal{H}$-harmonic and $\mathcal{H}$-subharmonic functions.

Corollary 4.1.3 (Invariant Mean-Value Property) *Let Ω be an open subset of $\mathbb{B}$. A real-valued C^2 function f is $\mathcal{H}$-subharmonic on Ω if and only if for all $a \in \Omega$,*

$$f(a) \leq \int_{\mathbb{S}} f(\varphi_a(rt))\,d\sigma(t) \tag{4.1.4}$$

for all $r > 0$ such that $E(a, r) \subset \Omega$. Furthermore, f is $\mathcal{H}$-harmonic on $\mathbb{B}$ if and only if equality holds in (4.1.4).

Proof. Suppose f is $\mathcal{H}$-subharmonic on Ω and $a \in \Omega$. Then $f \circ \varphi_a$ is $\mathcal{H}$-subharmonic on $\varphi_a(\Omega)$. Since $\Delta_h(f \circ \varphi_a) \geq 0$ on $\varphi_a(\Omega)$, by part (b) of the previous theorem we have

$$f(a) = (f \circ \varphi_a)(0) \leq \int_{\mathbb{S}} (f \circ \varphi_a)(rt)\,d\sigma(t)$$

for all $r > 0$ such that $B(0, r) \subset \varphi_a(\Omega)$. Clearly, if f is $\mathcal{H}$-harmonic on $\mathbb{B}$, then equality holds in (4.1.4).

Conversely, suppose $f \in C^2(\Omega)$ satisfies (4.1.4) for all $a \in \Omega$ and $r > 0$ such that $E(a, r) \subset \Omega$. Let $a \in \Omega$ be arbitrary, and set $h(x) = (f \circ \varphi_a)(x)$. Then there exists $r_o > 0$ such that

$$h(0) \leq \int_{\mathbb{S}} h(rt)d\sigma(t)$$

for all r, $0 < r < r_o$. Since h is C^2 in a neighborhood of 0, as a consequence of the Taylor expansion of h about 0,

$$\int_{\mathbb{S}} \{h(rt) - h(0)\}\, d\sigma(t) = \frac{r^2}{2n}\Delta h(0) + O(r^3). \tag{4.1.5}$$

Therefore,

$$\Delta h(0) = \lim_{r\to 0} \frac{2n}{r^2} \int_{\mathbb{S}} \{h(rt) - h(0)\}\, d\sigma(t),$$

and thus $\Delta_h f(a) = \Delta h(0) \geq 0$. Hence f is $\mathcal{H}$-subharmonic on $\mathbb{B}$. □

Remarks 4.1.4 **(a)** *If $f \in C^2(\mathbb{B})$ is $\mathcal{H}$-subharmonic on $\mathbb{B}$, then inequality (4.1.4) holds for all r, $0 < r < 1$. Furthermore, as a consequence of Theorem 4.1.1 the integral mean*

$$M(f,r) = \int_{\mathbb{S}} f(rt)d\sigma(t)$$

is a non-decreasing function of r, $0 < r < 1$.

(b) *As a consequence of the above proof, for $f \in C^2(\Omega)$, we have*

$$\Delta_h f(a) = \lim_{r\to 0} \frac{2n}{r^2} \int_{\mathbb{S}} \{f(\varphi_a(rt)) - f(a)\}d\sigma(t). \tag{4.1.6}$$

One other consequence of Theorem 4.1.1 is the following corollary.

Corollary 4.1.5 *If $f \in C_c^2(\mathbb{B})$, then for all $a \in \mathbb{B}$,*

$$f(a) = -\int_{\mathbb{B}} G_h(a,x)\Delta_h f(x)d\tau(x),$$

where G_h is the hyperbolic Green's function for $\mathbb{B}$.

Proof. Since f has compact support in $\mathbb{B}$, letting $r \to 1$ in Theorem 4.1.1(b) gives

$$f(0) = -\int_{\mathbb{B}} g_h(|x|)\Delta_h f(x)d\tau(x).$$

Applying this to $f \circ \varphi_a$ gives the desired result. □

We conclude this section with the following version of the maximum principle.

Theorem 4.1.6 (Maximum Principle) *Suppose Ω is an open subset of $\mathbb{B}$ and that $f \in C^2(\Omega)$ is $\mathcal{H}$-subharmonic in Ω and continuous on $\overline{\Omega}$. If $f \leq 0$ on $\partial\Omega$, then $f \leq 0$ in Ω.*

Proof. Set $h(x) = f(x) + \epsilon|x|^2$. Then $h \le \epsilon$ on $\partial\Omega$ and for all $x \in \Omega$,

$$\Delta_h h(x) = \Delta_h f(x) + 2\epsilon[n(1-|x|^2)^2 + 2(n-2)|x|^2(1-|x|^2)].$$

Thus $\Delta_h h(x) > 0$ for all $x \in \Omega$. If h has a local maximum at some point $x \in \Omega$, then $h \circ \varphi_x$ has a local maximum at 0. This, however, is impossible since

$$\Delta(h \circ \varphi_x)(0) = \Delta_h h(x) > 0.$$

Thus $h(x) < \epsilon$ for all $x \in \Omega$. Finally, since $f(x) \le h(x)$ for all x, letting $\epsilon \to 0$ gives $f(x) \le 0$ on Ω. □

4.2 The Special Case $n = 2$

In this section we consider some of the previous results in the setting of the unit disc $\mathbb{D}$ in the complex plane $\mathbb{C}$. When $n = 2$ with $z = x + iy$,

$$\Delta_h f(z) = (1-|z|^2)^2 \Delta f(z) = 4(1-|z|^2)^2 \frac{\partial^2 f(z)}{\partial z \partial \overline{z}}.$$

Thus f is $\mathcal{H}$-subharmonic ($\mathcal{H}$-harmonic) on $\mathbb{D}$ if and only if f is Euclidean subharmonic (Euclidean harmonic) on $\mathbb{D}$.

When $n = 2$, the Möbius transformation $\varphi_w(z)$ becomes

$$\varphi_w(z) = \frac{w - z}{1 - \overline{w}z},$$

and the Green's function $G_h(z, w)$ and invariant measure τ are

$$G_h(z,w) = \frac{1}{2}\log\left|\frac{1-\overline{w}z}{w-z}\right| \quad \text{and} \quad d\tau(z) = \frac{1}{\pi}(1-|z|^2)^{-2}dA(z),$$

where dA denotes area measure in $\mathbb{D}$. Thus Corollary 4.1.5 becomes

$$f(z) = -\frac{1}{2\pi}\int_{\mathbb{D}} \log\left|\frac{1-\overline{w}z}{w-z}\right| \Delta f(w)\, dA(w)$$

for all $f \in C^2(\mathbb{D})$ with compact support.

Suppose now that f is subharmonic on $\mathbb{D}$. With $\varphi_a(z) = (a-z)/(1-\overline{a}z)$, the mean-value property 4.1.4 becomes

$$f(a) \le \frac{1}{2\pi}\int_0^{2\pi} f(\varphi_a(re^{it}))dt \tag{4.2.1}$$

for all $r, 0 < r < 1$. Here, the integral is the average of f over the non-Euclidean circle $\{\varphi_a(re^{it}) \colon\ 0 \le t < 2\pi\}$ with non-Euclidean center a. In contrast, the Euclidean mean-value property is that

$$f(a) \le \frac{1}{2\pi}\int_0^{2\pi} f(a + re^{it})dt, \tag{4.2.2}$$

valid for all r such that $B(a,r) \subset \mathbb{D}$. Here the integral is over the Euclidean circle $\{a + re^{it} : 0 \le t < 2\pi\}$. If f is harmonic, then equality holds in both (4.2.1) and (4.2.2).

Suppose now that f is subharmonic and continuous on $\overline{\mathbb{D}}$. Since $f \circ \varphi_a$ is also continuous on $\overline{\mathbb{D}}$, letting $r \to 1$ in (4.2.1) gives

$$f(a) \le \frac{1}{2\pi}\int_0^{2\pi} f(\varphi_a(e^{it}))dt = \frac{1}{2\pi i}\int_{|\zeta|=1} \frac{f(\varphi_a(\zeta))}{\zeta}d\zeta.$$

Letting $w = \varphi_a(\zeta)$ we have $\zeta = \varphi_a(w)$. Thus $d\zeta = \varphi_a'(w)dw$ and

$$\begin{aligned}\frac{1}{2\pi i}\int_{|\zeta|=1}\frac{f(\varphi_a(\zeta))}{\zeta}d\zeta &= \frac{1}{2\pi i}\int_{|w|=1} f(w)\frac{\varphi_a'(w)}{\varphi_a(w)}dw \\ &= \frac{1}{2\pi}\int_0^{2\pi} f(e^{it})\frac{\varphi_a'(e^{it})e^{it}}{\varphi_a(e^{it})}dt.\end{aligned}$$

But

$$\frac{\varphi_a'(e^{it})e^{it}}{\varphi_a(e^{it})} = \frac{-(1-|a|^2)e^{it}}{(a-e^{it})(1-\bar{a}e^{it})} = \frac{(1-|a|^2)}{|a-e^{it}|^2}.$$

Therefore

$$f(a) \le \frac{1}{2\pi}\int_0^{2\pi} P(a,e^{it})f(e^{it})dt, \tag{4.2.3}$$

where

$$P(a,e^{it}) = \frac{1-|a|^2}{|a-e^{it}|^2}$$

is the Poisson kernel on $\mathbb{D}$. If f is harmonic on $\mathbb{D}$, then equality holds in the above and (4.2.3) is the Poisson integral formula for f. In Section 5.3 we will establish the Poisson integral formula for $\mathcal{H}$-harmonic functions on $\mathbb{B}$.

Suppose now that f is harmonic on $\mathbb{D}$. When $n = 2$, the function $f_r(z) = f(rz)$ is also harmonic on $\mathbb{D}$ for all r, $0 < r < 1$. Applying (4.2.3) to f_r we have

$$f_r(a) = \frac{1}{2\pi}\int_0^{2\pi} P(a,e^{it})f_r(e^{it})dt.$$

For $w \in \mathbb{C}$ with $|w| < r$, set $a = w/r$. Then

$$f(w) = \frac{1}{2\pi}\int_0^{2\pi} \frac{r^2 - |w|^2}{|w - re^{it}|^2}f(re^{it})dt, \tag{4.2.4}$$

which is the classical Poisson integral formula for the disc $|w| < r$.

If f is C^2, then as above f is $\mathcal{H}$-subharmonic on $\mathbb{D}$ if and only if f is subharmonic on $\mathbb{D}$. As in the proof of (4.2.4), we then have

$$f(w) \le \frac{1}{2\pi}\int_0^{2\pi} \frac{r^2 - |w|^2}{|w - re^{it}|^2} f(re^{it})dt$$

for all w, $|w| < r$. In this case, the function

$$H(w) = \frac{1}{2\pi}\int_0^{2\pi} \frac{r^2 - |w|^2}{|w - re^{it}|^2} f(re^{it})dt$$

is harmonic in $|w| < r$, continuous on $|w| \le r$ with boundary values $f(re^{it})$.

An excellent reference for potential theory in the complex plane is the monograph by T. Ransford [69].

4.3 $\mathcal{H}$-Subharmonic Functions

We now extend our definition of $\mathcal{H}$-subharmonic functions to the class of upper semicontinuous functions. We begin with a brief review of upper semicontinuous functions.

Definition 4.3.1 *Let D be a subset of $\mathbb{R}^n$. A function $f : D \to [-\infty, \infty)$ is* **upper semicontinuous** *at $x_o \in D$ if for every $\alpha \in \mathbb{R}$ with $\alpha > f(x_o)$ there exists a $\delta > 0$ such that*

$$f(x) < \alpha \quad \textit{for all} \quad x \in D \cap B(x_o, \delta).$$

The function f is upper semicontinuous on D if f is upper semicontinuous at each point of D. A function $g : D \to (-\infty, \infty]$ is **lower semicontinuous** *if $-g$ is upper semicontinuous.*

If $f(x_o) > -\infty$, then the above definition is equivalent to given $\epsilon > 0$, there exists a $\delta > 0$ such that

$$f(x) < f(x_o) + \epsilon \quad \text{for all} \quad x \in D \cap B(x_o, \delta).$$

From the definition it follows that f is upper semicontinuous on D if and only if $\{x \in D : f(x) < \alpha\}$ is open in D for each $\alpha \in \mathbb{R}$. Characteristic functions of closed sets are upper semicontinuous whereas characteristic functions of open sets are lower semicontinuous. It is an easy exercise to show that f is upper semicontinuous at x_o if and only if

$$\limsup_{x \to x_o} f(x) \le f(x_o).$$

The following theorem provides some elementary properties of upper semicontinuous functions.

Theorem 4.3.2 *Let S be a non-empty subset of* $\mathbb{R}^n$.

(a) *If f is upper semicontinuous on S and K is a compact subset of S, then*

$$\sup_{x\in K} f(x) < \infty$$

and the supremum is assumed at some $x_o \in K$.

(b) *If* $f_1, \ldots, f_n$ *are upper semicontinuous on S, then*

$$f(x) = \max\{f_1(x), \ldots, f_n(x)\}$$

is upper semicontinuous on S.

(c) *If* $\{f_\alpha\}_{\alpha\in A}$ *are upper semicontinuous on S, then*

$$f(x) = \inf_{\alpha\in A}\{f_\alpha(x)\}$$

is upper semicontinuous on S.

(d) *If f is upper semicontinuous and bounded above on S, then there exists a decreasing sequence* $\{f_n\}$ *of continuous real-valued functions on S, bounded above, with*

$$f(x) = \lim_{n\to\infty} f_n(x) \quad \textit{for all} \quad x \in S.$$

Proof. For the proof of (a) let $\alpha = \sup_{x\in K} f(x)$, and let $\{x_n\}$ be a sequence in K such that $f(x_n) \to \alpha$. By compactness there exists a subsequence $\{x_{n_k}\}$ of $\{x_n\}$ that converges to some $x_o \in K$. Since $f(x_o)$ is finite, by upper semicontinuity of the function f, given $\epsilon > 0$ there exists a $\delta > 0$ such that

$$f(x) < f(x_o) + \epsilon$$

for all $x \in B(x, \delta) \cap S$. In particular, there exists k_o such that $f(x_{n_k}) < f(x_o) + \epsilon$ for all $k \geq k_o$ Thus

$$\alpha \leq f(x_o) + \epsilon \leq \alpha + \epsilon.$$

Since $\epsilon > 0$ was arbitrary, we have $f(x_o) = \alpha$.

(d) Let $M \in \mathbb{R}$ be such that $f(x) \leq M$ for all $x \in S$. For each positive integer n define

$$f_n(x) = \sup_{w\in S}\{f(w) - n|x - w|\}, \quad x \in S.$$

Then $f_n(x) \leq M$ for all n, $f_n(x) \geq f(x)$, and $f_n(x) \geq f_m(x)$ for all positive integers m, n, with $m \geq n$. Hence it remains only to be shown that each f_n is continuous on S and that $f_n(x) \to f(x)$ for all $x \in S$.

Fix n and let $\epsilon > 0$ be given. If $x_1, x_2 \in S$ with $|x_1 - x_2| < \epsilon/n$, then

$$|x_2 - w| < |x_1 - w| + \frac{1}{n}\epsilon,$$

and thus

$$f(w) - n|x_1 - w| < f(w) - n|x_2 - w| + \epsilon.$$

Therefore, $f_n(x_1) < f_n(x_2) + \epsilon$. By symmetry we have $|f_n(x_1) - f_n(x_2)| < \epsilon$, and thus f_n is continuous (in fact uniformly continuous) on S.

We next prove that $f_n(x) \to f(x)$ pointwise on S. Fix $x_o \in S$ and suppose $f(x_o) > -\infty$. Without loss of generality we can assume that $\sup_w f(w) \leq 1$ and $f(x_o) = 0$. Let $\epsilon > 0$ be given. Since f is upper semicontinuous there exists $\delta > 0$ such that $f(w) < \epsilon$ for all $w \in S$ with $|x_o - w| < \delta$. If $|x_o - w| \geq \delta$, then for all n with $n > 1/\delta$,

$$f(w) - n|x_o - w| \leq 1 - 1 = 0.$$

If $|x_o - w| < \delta$, then $f(w) - n|x_o - w| < \epsilon$ for all n. Hence for all $n > 1/\delta$,

$$0 = f(x_o) \leq f_n(x_o) = \sup_w\{f(w) - n|x_o - w|\} < \epsilon.$$

Thus $f_n(x_o) \to f(x_o)$. If $f(x_o) = -\infty$, then a simple modification of the above shows that $f_n(x_o) \to -\infty$.

The proofs of (b) and (c) are straightforward and are omitted. □

Definition 4.3.3 *Let Ω be an open subset of $\mathbb{B}$. An upper semicontinuous function $f : \Omega \mapsto [-\infty, \infty)$, with $f \not\equiv -\infty$, is $\mathcal{H}$-**subharmonic** on Ω if*

$$f(a) \leq \int_{\mathbb{S}} f(\varphi_a(rt))\, d\sigma(t) \tag{4.3.1}$$

*for all $a \in \Omega$ and all r sufficiently small. A function f is $\mathcal{H}$-**superharmonic** if $-f$ is $\mathcal{H}$-subharmonic.*

Inequality (4.3.1) is the Möbius invariant mean-value inequality. Definition 4.3.3 is of course equivalent to Definition 4.0.1 for the class of C^2 functions. In the following theorem we prove that with this definition the class of $\mathcal{H}$-subharmonic functions is again invariant under $\mathcal{M}$.

Theorem 4.3.4 *If f is $\mathcal{H}$-subharmonic on $\mathbb{B}$, then $f \circ \psi$ is $\mathcal{H}$-subharmonic for all $\psi \in \mathcal{M}(\mathbb{B})$.*

Proof. For $\psi \in \mathcal{M}$ and $a \in \mathbb{B}$, let $b = \psi(a)$. Then $(\varphi_b \circ \psi \circ \varphi_a)(0) = 0$, and thus $\psi(\varphi_a(x)) = \varphi_b(Ax)$ for some $A \in O(n)$. Consequently, by the $O(n)$ invariance of σ,

$$\begin{aligned}\int_{\mathbb{S}}(f\circ\psi)(\varphi_a(rt))\,d\sigma(t) &= \int_{\mathbb{S}} f(\varphi_b(Art))\,d\sigma(t)\\ &= \int_{\mathbb{S}} f(\varphi_b(rt))\,d\sigma(t) \geq f(b) = (f\circ\psi)(a).\end{aligned}$$

□

There is also a volume version of the invariant mean-value inequality.

Theorem 4.3.5 *Let Ω be an open subset of $\mathbb{B}$. If f is $\mathcal{H}$-subharmonic on Ω, then for each $a \in \Omega$,*

$$f(a) \leq \frac{1}{\tau(B_r)}\int_{E(a,r)} f(x)d\tau(x) \tag{4.3.2}$$

for all r sufficiently small such that $E(a,r) \subset \Omega$. If f is $\mathcal{H}$-harmonic on Ω, then equality holds in (4.3.2).

Proof. The inequality follows from the $\mathcal{M}$-invariance of τ, and is obtained by multiplying both sides of (4.3.1) by $n\rho^{n-1}(1-\rho^2)^{-n}$ and integrating from 0 to r. □

Remark 4.3.6 *If f is $\mathcal{H}$-harmonic on $\mathbb{B}$, then*

$$f(\psi(0)) = \int_{\mathbb{S}} f(\psi(rt))\,d\sigma(t)$$

for all $\psi \in \mathcal{M}(\mathbb{B})$. By multiplying by $n\rho^{n-1}$ and integrating we also have

$$f(\psi(0)) = \int_{\mathbb{B}} f(\psi(w))\,d\nu(w) \tag{4.3.3}$$

for all $\psi \in \mathcal{M}(\mathbb{B})$. This leads to the following question.

Question: *If $f \in L^1(\mathbb{B})$ satisfies (4.3.3) for every $\psi \in \mathcal{M}(\mathbb{B})$, is f $\mathcal{H}$-harmonic?*

It is known that the answer is yes when $n = 2$. On the Hermitian ball $\mathbb{B}$ in $\mathbb{C}^m$ the answer to the analogous question is very surprising. In [3] P. Ahern, M. Flores, and W. Rudin answered the question in the affirmative when $m \leq 11$, and in the negative when $m \geq 12$.

For C^2 $\mathcal{H}$-subharmonic functions f there is also an invariant volume version of (4.1.6). If we set $h(x) = (f\circ\varphi_a)(x)$ and integrate (4.1.5), then for all r sufficiently small,

$$\int_{B_r}\{h(x)-h(0)\}\,d\tau(x) = \tfrac{1}{2}\Delta h(0)c(r) + O(r^{n+3}),$$

where

$$c(r) = \int_0^r \rho^{n+1}(1-\rho^2)^{-n} d\rho.$$

By L'Hospital's rule,

$$\lim_{r\to 0} \frac{r^2\tau(B_r)}{c(r)} = n + 2 \lim_{r\to 0}(1-r^2)\left\{\frac{\tau(B_r)}{r^n/(1-r^2)^{n-1}}\right\},$$

which by (3.3.6)

$$= n + 2.$$

Also, since $r^2\tau(B_r) \approx r^{n+2}$ for r sufficiently small, we have

$$\Delta_h f(a) = \lim_{r\to 0} \frac{2(n+2)}{r^2\tau(B_r)} \int_{E(a,r)} \{f(x) - f(a)\}\, d\tau(x). \tag{4.3.4}$$

We now express identity (4.3.4) in terms of the invariant convolution given in Definition 3.4.1. If we define Ω_r by

$$\Omega_r(x) = \begin{cases} \dfrac{1}{\tau(B_r)}, & |x| \le r, \\ 0, & |x| > r, \end{cases}$$

then

$$(f * \Omega_r)(a) - f(a) = \frac{1}{\tau(B_r)} \int_{E(a,r)} \{f(x) - f(a)\}\, d\tau(x).$$

Thus by (4.3.4), if $f \in C^2(\mathbb{B})$, then

$$(\Delta_h f)(a) = \lim_{r\to 0} \frac{2(n+2)}{r^2} \left[(f * \Omega_r)(a) - f(a)\right]. \tag{4.3.5}$$

If f has compact support, then the convergence is uniform on $\mathbb{B}$.

4.4 Properties of $\mathcal{H}$-Subharmonic Functions

In this section we prove several properties of $\mathcal{H}$-subharmonic functions. Most of these are an immediate consequence of the definition.

Theorem 4.4.1 **(a)** *If u_1, u_2 are $\mathcal{H}$-subharmonic on $\mathbb{B}$, then*

$$u(x) = \max\{u_1(x), u_2(x)\}$$

is $\mathcal{H}$-subharmonic on $\mathbb{B}$.

(b) *If u is $\mathcal{H}$-subharmonic on $\mathbb{B}$ and $\varphi : [-\infty, \infty)$ is an increasing convex function, then $\varphi \circ u$ is $\mathcal{H}$-subharmonic on $\mathbb{B}$.*

(c) *If u_n is a decreasing sequence of $\mathcal{H}$-subharmonic functions, then*

$$u(x) = \lim_{n\to\infty} u_n(x)$$

is either identically $-\infty$ or $\mathcal{H}$-subharmonic on $\mathbb{B}$.

Proof. The proof of (a) is straightforward and is left as an exercise. The proof of (b) is an immediate consequence of Jensen's inequality. Recall that a function $\varphi : (a,b) \to \mathbb{R}$ is **convex** on (a,b) if whenever $x_1, x_2 \in (a,b)$ we have

$$\varphi(tx_1 + (1-t)x_2) \le t\varphi(x_1) + (1-t)\varphi(x_2)$$

for all $t \in [0,1]$. It is easily shown that a C^2 function φ is convex if and only if $\varphi''(x) \ge 0$. **Jensen's inequality** is as follows: if (X,μ) is a measure space with $\mu(X) = 1$ and φ is a convex function on $(-\infty,\infty)$, then

$$\varphi\left(\int_X f\,d\mu\right) \le \int_X \varphi\circ f\,d\mu$$

whenever f is an integrable function on X (see [71, Proposition 20]).

(c) Assume that $u \not\equiv -\infty$. For each $\alpha \in \mathbb{R}$ we have

$$\{x : u(x) < \alpha\} = \bigcup_{n=1}^{\infty}\{x : u_n(x) < \alpha\}.$$

Hence u is upper semicontinuous on $\mathbb{B}$. Also, for each $n \in \mathbb{N}$ and $a \in \mathbb{B}$, we have

$$u_n(a) \le \int_{\mathbb{S}} u_n(\varphi_a(rt))\,d\sigma(t).$$

Hence by the monotone convergence theorem,

$$u(a) \le \int_{\mathbb{S}} u(\varphi_a(rt))d\sigma(t).$$

Thus u is $\mathcal{H}$-subharmonic on $\mathbb{B}$. □

As a consequence of the previous theorem, if f is $\mathcal{H}$-harmonic on $\mathbb{B}$, then $|f|^p$ is $\mathcal{H}$-subharmonic for all p, $1 \le p < \infty$. Likewise, if f is $\mathcal{H}$-subharmonic on $\mathbb{B}$, then e^{cf}, $c > 0$, and $f^+(x) = \max\{f(x), 0\}$ are $\mathcal{H}$-subharmonic on $\mathbb{B}$.

Since many of the properties of subharmonic functions follow from the Euclidean analogue of (4.3.1) or (4.3.2), those same properties are still valid for $\mathcal{H}$-subharmonic functions on $\mathbb{B}$. In particular one has the following versions of the maximum principle.

Theorem 4.4.2 (Maximum Principle) *Let Ω be an open connected subset of $\mathbb{B}$.*

(a) (Version I) *If f is a non-constant $\mathcal{H}$-subharmonic function on Ω, then*

$$f(x) < \sup_{y\in\Omega} f(y) \qquad \textit{for all} \quad x \in \Omega.$$

(b) (Version II) *If f is $\mathcal{H}$-subharmonic on Ω and*

$$\limsup_{\substack{x\to\zeta \\ x\in\Omega}} f(x) \le M \qquad \textit{for all} \quad \zeta \in \partial\Omega,$$

then $f(x) \le M$ for all $x \in \Omega$.

Proof. (a) Let $M = \sup_{x\in\Omega} f(x)$. If $M = \infty$ then the result is certainly true. Hence assume $M < \infty$ and let $E = \{y \in \Omega : f(y) = M\}$. Since $f(y) \le M$ for all y, $E = \Omega \setminus \{y : f(y) < M\}$ and thus E is closed. We now show that E is open. Suppose $E \ne \phi$. Let $a \in E$ and let $r_a > 0$ be such that $E(a, r_a) \subset \Omega$ and that (4.3.2) holds for all r with $0 < r < r_a$. Then

$$M = f(a) \le \frac{1}{\tau(E(a,r))} \int_{E(a,r)} f(x)d\tau(x),$$

and thus

$$\int_{E(a,r)} [f(x) - M]\, d\tau(x) = 0.$$

Since $f(x) \le M$, the above implies that $f(x) = M$ a.e. on $E(a, r)$. But since f is upper semicontinuous, $f(x) = M$ for all $x \in E(a, r)$. Therefore $E(a, r) \subset E$ and thus E is open. Since Ω is connected, $E = \Omega$ or $E = \phi$. Finally, since f is non-constant, $E = \phi$, which proves the result.

(b) Let $\alpha = \sup_{x\in\Omega} f(x)$, and let $\{x_n\}$ be a sequence in Ω such that $f(x_n) \to \alpha$. Since $\{x_n\}$ is bounded, there exists a subsequence $\{x_{n_k}\}$ of $\{x_n\}$ that converges to some $x_o \in \overline{\Omega}$. But by part (a), $x_o \in \partial\Omega$. But then $\alpha = \lim_{k\to\infty} f(x_{n_k}) \le \limsup_{x\to x_o} f(x) \le M$. □

Another consequence of the mean-value inequality for $\mathcal{H}$-subharmonic functions is local integrability. As in Section 3.4, a measurable function f defined on an open subset Ω of $\mathbb{R}^n$ is **locally integrable** if

$$\int_K |f(x)|d\nu(x) < \infty$$

for every compact subset K of Ω. The set of locally integrable functions on Ω is denoted by $L^1_{loc}(\Omega)$.

Theorem 4.4.3 *Let Ω be an open, connected subset of $\mathbb{B}$. If f is $\mathcal{H}$-subharmonic on Ω, then f is locally integrable on Ω and $\{x \in \Omega : f(x) = -\infty\}$ has measure zero.*

Proof. By compactness it suffices to show that f is integrable on every hyperbolic ball $E(a,r)$ with $\overline{E(a,r)} \subset \Omega$. Set

$$U = \left\{x \in \Omega : \int_{E(x,r)} |f|\,d\tau < \infty \text{ for some } r > 0\right\},$$

and let

$$P = \{x \in \Omega : f(x) = -\infty\}.$$

If $a \in U$ and $r > 0$ is such that $\int_{E(a,r)} |f|\,d\tau < \infty$, then for every $y \in E(a,r)$,

$$\int_{E(y,\delta)} |f|\,d\tau < \infty$$

for every δ, $0 < \delta < r - |\varphi_a(x)|$. Thus U is an open subset of Ω. We next show that U is also non-empty.

Since f is not identically $-\infty$, $\Omega \setminus P \neq \phi$. Let $a \in \Omega \setminus P$ and choose $r > 0$ such that $\overline{E(a,r)} \subset \Omega$. Since $f(a) > -\infty$, by inequality (4.3.2)

$$\int_{E(a,r)} f(x)\,d\tau(x) > -\infty.$$

Let $M = \sup\{f(x) : x \in \overline{E(a,r)}\}$, which is finite since f is upper semicontinuous. Since

$$|f(x)| \leq |M| + M - f(x)$$

for all $x \in E(a,r)$, we have $\int_{E(a,r)} |f|\,d\tau < \infty$. Therefore U is non-empty.

We now show that U is also closed. Let $a \in \Omega \setminus U$, and let $r > 0$ be such that $\overline{E(a,3r)} \subset \Omega$. Since $a \in \Omega \setminus U$ we have

$$\int_{E(a,r)} f(x)\,d\tau(x) = -\infty.$$

Let $M < \infty$ be such that $f(x) \leq M$ for all $x \in E(a,3r)$. Then $M - f(x) \geq 0$ for all $x \in E(a,3r)$. If $x \in E(a,r)$, then by the triangle inequality for the pseudo-hyperbolic metric, $E(a,r) \subset E(x,2r)$. Therefore,

$$\int_{E(x,2r)} [M - f]\,d\tau \geq \int_{E(a,r)} [M - f]\,d\tau = \infty.$$

As a consequence

$$\int_{E(x,2r)} f\,d\tau = -\infty$$

for all $x \in E(a,r)$. Thus $\Omega \setminus U$ is open; that is, U is closed. Since Ω is connected and $U \neq \phi$, we have $U = \Omega$. Thus $f(x) > -\infty$ τ - a.e. and $\tau(P) = 0$. □

4.5 Approximation by C^∞ $\mathcal{H}$-Subharmonic Functions

We begin this section with the definition of an approximate identity.

Definition 4.5.1 *A sequence of non-negative continuous functions* $\{h_j\}$ *is an* **approximate identity** *in* $L^1(\mathbb{B}, \tau)$ *if*

(a) $\displaystyle\int_{\mathbb{B}} h_j d\tau = 1$, *and*

(b) $\displaystyle\lim_{j\to\infty} \int_{\mathbb{B}\setminus B(0,\delta)} h_j d\tau = 0$ *for every* $\delta > 0$.

Definition 4.5.2 *Let* $\{r_k\}$ *be a decreasing sequence with* $r_k \to 0$ *as* $k \to \infty$. *For each* k, *let* χ_k *be a non-negative* C^∞ *radial function on* $\mathbb{B}$ *with support contained in* $\{x : r_{k+1} < |x| < r_k\}$ *satisfying*

$$\int_{\mathbb{B}} \chi_k(x)\, d\tau(x) = 1.$$

It is easily seen that the sequence $\{\chi_k\}_{k=1}^{\infty}$ is a C^∞ approximate identity for $L^1(\mathbb{B}, \tau)$.

The following lemma shows the importance of an approximate identity, also sometimes referred to as an approximate unit.

Lemma 4.5.3 *Let* $\{\chi_k\}$ *be defined as above. Then*

$$\lim_{k\to\infty} (h * \chi_k) = h,$$

uniformly on $\mathbb{B}$ *if* $h \in C_c(\mathbb{B})$, *and locally in* L^p *if* $h \in L^p_{loc}(\mathbb{B})$, $1 \le p < \infty$.

Proof. Suppose $h \in C_c(\mathbb{B})$ with $K = \operatorname{supp} h$. Since K is compact, h is uniformly continuous on K and hence on $\mathbb{B}$. Thus given $\epsilon > 0$ there exists a $\delta > 0$ such that

$$|h(x) - h(y)| < \epsilon$$

for all $x, y \in \mathbb{B}$ with $y \in E(x, \delta)$. For each k let $h_k = h * \chi_k$ where $*$ is given in Definition 3.4.1. Since

$$h_k(x) = \int_{\mathbb{B}} h(y)\chi_k(\phi_x(y))d\tau(y),$$

we have

$$\begin{aligned}
|h(x) - h_k(x)| &\le \int_{\mathbb{B}} |h(x) - h(y)|\chi_k(\varphi_x(y))d\tau(y) \\
&\le \int_{E(x,\delta)} |h(x) - h(y)|\chi_k(\varphi_x(y))d\tau(y) + 2\|h\|_\infty \int_{\mathbb{B}\setminus E(x,\delta)} \chi_k(\varphi_x(y))d\tau(y) \\
&< \epsilon + 2\|h\|_\infty \int_{\mathbb{B}\setminus B(0,\delta)} \chi_k(y)d\tau(y).
\end{aligned}$$

Hence since $\{\chi_k\}$ is an approximate identity, $\lim_{k\to\infty} h_k(x) = h(x)$ uniformly on $\mathbb{B}$.

Suppose $h \in L^p_{loc}(\mathbb{B})$, $1 \le p < \infty$. Let K be a compact subset of $\mathbb{B}$. Redefine h on $\mathbb{B}$ such that $h(x) = 0$ for all $x \notin K$. For $\epsilon > 0$, choose $g \in C_c(\mathbb{B})$ with $\operatorname{supp} g \subset K$ and

$$\int_K |h-g|^p d\tau = \int_{\mathbb{B}} |h-g|^p d\tau < \epsilon^p.$$

Let $g_k = g * \chi_k$. Then by Minkowski's inequality,

$$\|h-h_k\|_p \le \|h-g\|_p + \|h_k - g_k\|_p + \|g_k - g\|_p.$$

Since g is continuous, $g_k \to g$ uniformly on $\mathbb{B}$. Furthermore, since $\operatorname{supp} g$ is compact and $\operatorname{supp}\chi_k \subset \{x : r_{k+1} < |x| < r_k\}$ we have that $\operatorname{supp} g_k$, as well as h_k, is also compact for sufficiently large k. Hence there exists k_o such that $\|g_k - g\|_p < \epsilon$ for all $k \ge k_o$. Thus for $k \ge k_o$ we have

$$\left[\int_K |h-h_k| d\tau\right]^{1/p} \le \|h-h_k\|_p \le 2\epsilon + \|h_k - g_k\|_p,$$

which by Theorem 3.4.2

$$\le 2\epsilon + \|h-g\|_p \|\chi_k\|_1 \le 3\epsilon,$$

from which the result now follows. □

We are now in a position to prove the following theorem.

Theorem 4.5.4 *Let $\{\chi_k\}$ be a C^∞ approximate identity as above. If f is $\mathcal{H}$-subharmonic on $\mathbb{B}$, then $\{f * \chi_k\}_{k=1}^\infty$ is a non-increasing sequence of C^∞ $\mathcal{H}$-subharmonic functions on $\mathbb{B}$ satisfying*

$$(f*\chi_k)(x) \ge f(x) \quad \text{and} \quad \lim_{k\to\infty} (f*\chi_k)(x) = f(x) \tag{4.5.1}$$

for all $x \in \mathbb{B}$.

Proof. Since χ_k is C^∞, the function $f * \chi_k$ is also C^∞. By integration in polar coordinates,

$$\begin{aligned}(f*\chi_k)(y) &= n\int_0^1 \frac{r^{n-1}}{(1-r^2)^n}\chi_k(r)\left[\int_{\mathbb{S}} f(\varphi_y(rt))\,d\sigma(t)\right] dr \qquad (4.5.2)\\ &\ge f(y)\int_{\mathbb{B}} \chi_k\,d\tau = f(y).\end{aligned}$$

Fix $a \in \mathbb{B}$, and let $\alpha > f(a)$. Since f is upper semicontinuous there exists $r > 0$ such that $f(w) < \alpha$ for all $w \in E(a, r)$. Hence if $r_k < r$,

$$(f * \chi_k)(a) = \int_{\mathbb{B}} f(w)\chi_k(\varphi_a(w))\, d\tau(w) \le \alpha \int_{\mathbb{B}} \chi_k\, d\tau = \alpha.$$

Thus $\limsup_{k\to\infty}(f * \chi_k)(a) \le f(a)$, which when combined with the above proves (4.5.1).

To show that $f * \chi_k$ is $\mathcal{H}$-subharmonic we use (4.3.5). Since χ_k and Ω_r are both radial, by Lemma 3.4.3

$$(f * \chi_k) * \Omega_r = (f * \Omega_r) * \chi_k.$$

But as above, by integration in polar coordinates $(f * \Omega_r) \ge f$. Therefore

$$(f * \chi_k) * \Omega_r = (f * \Omega_r) * \chi_k \ge f * \chi_k.$$

Thus by (4.3.5) $\Delta_h(f * \chi_k) \ge 0$. Therefore $f * \chi_k$ is $\mathcal{H}$-subharmonic.

Finally it remains to be shown that the sequence $\{f * \chi_k\}$ is non-increasing. For the proof we require that an $\mathcal{H}$-subharmonic function satisfies

$$\int_{\mathbb{S}} f(rt)\, d\sigma(t) \le \int_{\mathbb{S}} f(\rho t)\, d\sigma(t) \tag{4.5.3}$$

whenever $0 < r < \rho < 1$. By Remark 4.1.4 following the proof of Corollary 4.1.3, inequality (4.5.3) is valid whenever f is a C^2 $\mathcal{H}$-subharmonic function. Thus since $f \le f * \chi_k$,

$$\int_{\mathbb{S}} f(rt)d\sigma(t) \le \int_{\mathbb{S}} (f * \chi_k)(rt)d\sigma(t) \le \int_{\mathbb{S}} (f * \chi_k)(\rho t)d\sigma(t).$$

But since $\mathcal{H}$-subharmonic functions are locally integrable and bounded above on compact sets, we have

$$\limsup_{k\to\infty} \int_{\mathbb{S}} (f * \chi_k)(\rho t)d\sigma(t) \le \int_{\mathbb{S}} f(\rho t)d\sigma(t),$$

which proves (4.5.3).

Since $(f * \chi_k)(a) = ((f \circ \varphi_a) * \chi_k)(0)$, it suffices to prove that the sequence $\{(f * \chi_k)(0)\}$ is non-increasing. Suppose $m > k$. Since the support of χ_k is contained in $\{r_{k+1} < |x| < r_k\}$ and $r_{k+1} \ge r_m$, we have

$$\begin{aligned}(f * \chi_k)(0) &= n\int_{r_{k+1}}^{r_k} r^{n-1}(1-r^2)^{-n}\chi_k(r)\int_{\mathbb{S}} f(rt)\, d\sigma(t)\, dr \\ &\ge \int_{\mathbb{S}} f(r_m t)\, d\sigma(t) \int_{\mathbb{B}} \chi_k\, d\tau\end{aligned}$$

$$
\begin{aligned}
&= \int_{\mathbb{S}} f(r_m t)\, d\sigma(t) \int_{\mathbb{B}} \chi_m \, d\tau \\
&\geq n \int_{r_{m+1}}^{r_m} r^{n-1}(1-r^2)^{-n} \chi_m(r) \int_{\mathbb{S}} f(rt)\, d\sigma(t)\, dr \\
&= (f * \chi_m)(0).
\end{aligned}
$$

□

Corollary 4.5.5 *If f is a continuous function on $\mathbb{B}$ that is both $\mathcal{H}$-subharmonic and $\mathcal{H}$-superharmonic on $\mathbb{B}$, then f is C^∞ on $\mathbb{B}$ and satisfies $\Delta_h f(x) = 0$ for all $x \in \mathbb{B}$.*

Proof. If f satisfies equality in (4.3.1), then by (4.5.2) $(f * \chi_k)(x) = f(x)$ for all $x \in \mathbb{B}$. Thus f is C^∞ on $\mathbb{B}$ and satisfies $\Delta_h f = 0$. □

4.6 The Weak Laplacian and Riesz Measure

Our goal in this section is to provide a characterization of $\mathcal{H}$-subharmonic functions in terms of the **weak Laplacian**. For this we require Green's identity (see [74]) for the invariant Laplacian. If $f, g \in C^2(\mathbb{B})$ and one of them has compact support, then

$$\int_{\mathbb{B}} f \Delta_h g \, d\tau = \int_{\mathbb{B}} g \Delta_h f \, d\tau. \tag{4.6.1}$$

Thus if f is a C^2 $\mathcal{H}$-subharmonic function, we have $\int_{\mathbb{B}} f \Delta_h \psi \, d\tau \geq 0$ for all $\psi \in C_c^2(\mathbb{B})$ with $\psi \geq 0$. Thus we make the following definition.

Definition 4.6.1 *For $f \in L^1_{loc}$ we say that $\Delta_h f \geq 0$ in the* **weak sense** *if*

$$\int_{\mathbb{B}} f \Delta_h \psi \, d\tau \geq 0$$

for all $\psi \in C_c^2(\mathbb{B})$ with $\psi \geq 0$.

Theorem 4.6.2 *If f is $\mathcal{H}$-subharmonic on $\mathbb{B}$, then*

$$\int_{\mathbb{B}} f(z) \Delta_h \psi(z) \, d\tau(z) \geq 0 \tag{4.6.2}$$

for all $\psi \in C_c^2(\mathbb{B})$ with $\psi \geq 0$. Conversely, if $f \in L^1_{loc}(\mathbb{B})$ is such that (4.6.2) holds for all $\psi \in C_c^2(\mathbb{B})$ with $\psi \geq 0$, then there exists an $\mathcal{H}$-subharmonic function F on $\mathbb{B}$ such that $F = f$ a.e. on $\mathbb{B}$. Furthermore, if $\int_{\mathbb{B}} f \Delta_h \psi \, d\tau = 0$, then there exists an $\mathcal{H}$-harmonic function F on $\mathbb{B}$ such that $F = f$ a.e. on $\mathbb{B}$.

Proof. Let $\{\chi_k\}$ be a C^∞ approximate identity as given in Definition 4.5.2. Suppose f is $\mathcal{H}$-subharmonic on $\mathbb{B}$. Set $f_k = f * \chi_k$. Then by Theorem 4.5.4, $\{f_k\}$ is a non-increasing sequence of C^∞ $\mathcal{H}$-subharmonic functions on $\mathbb{B}$ that converges to f everywhere on $\mathbb{B}$. Thus by Green's identity (4.6.1) and the monotone convergence theorem,

$$\begin{aligned}\int_{\mathbb{B}} f\,\Delta_h\psi\,d\tau &= \lim_{k\to\infty}\int_{\mathbb{B}} f_k\,\Delta_h\psi\,d\tau\\ &= \lim_{k\to\infty}\int_{\mathbb{B}} \psi\,\Delta_h f_k\,d\tau \geq 0\end{aligned}$$

for all $\psi \in C_c^2(\mathbb{B})$ with $\psi \geq 0$. This proves (4.6.2).

Conversely, suppose $f \in L^1_{loc}(\mathbb{B})$ satisfies (4.6.2). Let f_k be defined as above. Since $\varphi_0(x) = -x$, if we set $J = -I$ (minus the identity on $\mathbb{B}$), then

$$(f * [(\Delta_h\psi)\circ J])(0) = \int_{\mathbb{B}} f(x)\Delta_h\psi(x)\,d\tau(x).$$

Thus in the notation of convolutions, hypothesis (4.6.2) is just

$$(f * [(\Delta_h\psi)\circ J])(0) \geq 0$$

for all $\psi \in C_c^2(\mathbb{B})$ with $\psi \geq 0$. It follows from the definition that

$$\Delta_h f_k(a) = \Delta_h(f * \chi_k)(a) = (f * [(\Delta_h(\chi_k\circ\varphi_a))\circ J])(0).$$

Thus since $\chi_k \circ \varphi_a$ is C^∞ with compact support, $\Delta_h f_k(a) \geq 0$. Therefore f_k is $\mathcal{H}$-subharmonic on $\mathbb{B}$.

We now show that the sequence $\{f_k\}$ is non-increasing. Suppose $k > m$ and j is arbitrary. Since f_j is $\mathcal{H}$-subharmonic and χ_m is radial, by Theorem 3.4.3 we have

$$\begin{aligned}f_m * \chi_j = (f * \chi_m) * \chi_j &= (f * \chi_j) * \chi_m\\ &\geq (f * \chi_j) * \chi_k = f_k * \chi_j.\end{aligned}$$

Since f_m, f_k are continuous, $\lim_{j\to\infty} f_m * \chi_j = f_m$, with a similar result for f_k. Thus the sequence $\{f_k\}$ is non-increasing. Define

$$F(x) = \lim_{k\to\infty} f_k(x),$$

which exists everywhere on $\mathbb{B}$. As a consequence of the mean-value property, F is either $\mathcal{H}$-subharmonic on $\mathbb{B}$ or $F \equiv -\infty$. But by Lemma 4.5.3, $\{f_k\}$ converges to f locally in L^1, and thus $F = f$ a.e. on $\mathbb{B}$. □

Theorem 4.6.3 *If f is $\mathcal{H}$-subharmonic on $\mathbb{B}$, then there exists a unique regular Borel measure μ_f on $\mathbb{B}$ such that*

$$\int_{\mathbb{B}} \psi \, d\mu_f = \int_{\mathbb{B}} f \Delta_h \psi \, d\tau \tag{4.6.3}$$

for all $\psi \in C_c^2(\mathbb{B})$.

Definition 4.6.4 *If f is $\mathcal{H}$-subharmonic on $\mathbb{B}$, the unique regular Borel measure μ_f satisfying (4.6.3) is called the* **Riesz measure** *of f.*

Proof. Let f be $\mathcal{H}$-subharmonic on $\mathbb{B}$. By (4.6.2),

$$L(\psi) = \int_{\mathbb{B}} f \, \Delta_h \psi \, d\tau$$

defines a non-negative linear functional on $C_c^\infty(\mathbb{B})$. We extend L to $C_c(\mathbb{B})$ as follows. Let $\psi \in C_c(\mathbb{B})$. Choose a sequence $\{\psi_k\} \subset C_c^\infty(\mathbb{B})$ such that $\psi_k \to \psi$ uniformly on $\mathbb{B}$. Choose a compact subset K of $\mathbb{B}$ such that the supports of ψ and ψ_k, $k = 1, 2, \ldots$ are contained in K. Let V be a relatively compact open subset of $\mathbb{B}$ such that $K \subset V$, and let $h \in C_c^\infty(\mathbb{B})$, $0 \le h \le 1$, be such that $h \equiv 1$ on K and the support of h is contained in V. Set

$$\epsilon_{k,m} = \sup_{x \in K} |\psi_k(x) - \psi_m(x)|.$$

Then for all $x \in \mathbb{B}$,

$$-\epsilon_{k,m} h(x) \le \psi_k(x) - \psi_m(x) \le \epsilon_{k,m} h(x).$$

Thus since L is positive,

$$|L(\psi_k) - L(\psi_m)| \le \epsilon_{k,m} \, L(h).$$

Therefore $\{L(\psi_k)\}$ is Cauchy. Define

$$L(\psi) = \lim_{k \to \infty} L(\psi_k).$$

It is easy to show that $L(\psi)$ is independent of the choice of $\{\psi_k\}$, and thus defines a non-negative linear functional on $C_c(\mathbb{B})$. The result now follows by the Riesz representation theorem for non-negative linear functionals on $C_c(\mathbb{B})$ ([39, Theorem 12.6]). □

4.7 Quasi-Nearly $\mathcal{H}$-Subharmonic Functions

Following the notation introduced by M. Pavlović and J. Riihentaus [66] we make the following definition.

Definition 4.7.1 *A non-negative locally integrable function g on $\mathbb{B}$ is said to be* **quasi-nearly $\mathcal{H}$-subharmonic** *if there exists a constant C such that*

$$g(a) \le \frac{C}{r^n} \int_{E(a,r)} g(x)d\tau(x)$$

for all $a \in \mathbb{B}$ and all r, $0 < r < r_o$ for some $r_o > 0$.

If f is a non-negative $\mathcal{H}$-subharmonic function, then so is f^p for all $p \ge 1$. Thus by Theorem 4.3.5

$$f^p(a) \le \frac{1}{\tau(B_r)} \int_{E(a,r)} f^p(y)d\tau(y).$$

But by (3.3.6)

$$\tau(B_r) = n \int_0^r \frac{\rho^{n-1}}{(1-\rho^2)^n} d\rho \ge n \int_0^r \rho^{n-1} d\rho = r^n.$$

Thus

$$f^p(a) \le \frac{1}{r^n} \int_{E(a,r)} f^p(y)d\tau(y).$$

Hence f^p is quasi-nearly $\mathcal{H}$-subharmonic for all $p \ge 1$. Furthermore, as a consequence of Theorem 4.7.3, if in addition f is continuous, then the result is also valid for all p, $0 < p < 1$. Thus if h is $\mathcal{H}$-harmonic on $\mathbb{B}$, then $|h|^p$ is quasi-nearly $\mathcal{H}$-subharmonic for all p, $0 < p < \infty$.

Remarks 4.7.2 **(i)** *As a consequence of Exercise 4.8.3, a non-negative locally integrable function g is quasi-nearly $\mathcal{H}$-subharmonic if and only if for each δ, $0 < \delta < \delta_o$, for some $\delta_o, 0 < \delta_o < \frac{1}{2}$, there exists a constant C_δ such that*

$$g(a) \le C_\delta \int_{B(a,c_\delta(1-|a|^2))} g(y)d\tau(y), \tag{4.7.1}$$

where $c_\delta = \delta/(1-\delta)$, or equivalently,

$$g(a) \le \frac{C_\delta}{(1-|a|^2)^n} \int_{B(a,c_\delta(1-|a|^2))} g(y)d\nu(y). \tag{4.7.2}$$

The last inequality follows from the fact that $(1-|y|^2) \approx (1-|a|^2)$ for all $y \in B(a, \delta(1-|a|^2))$. Also, for $0 < \delta < \frac{1}{2}$, $B(a, c_\delta(1-|a|^2)) \subset B(a, 2\delta(1-|a|^2))$.

(ii) *A non-negative locally integrable function g is said to be (Euclidean) quasi-nearly subharmonic if*

$$g(a) \leq \frac{C}{r^n} \int_{B(a,r)} g(y)d\nu(y) \tag{4.7.3}$$

for all $r > 0$ such that $B(a,r) \subset \mathbb{B}$. If we take $r = \delta(1 - |a|^2)$ for fixed δ, $0 < \delta < \frac{1}{2}$, then

$$g(a) \leq \frac{C_\delta}{(1 - |a|^2)^n} \int_{B(a,\delta(1-|a|^2))} g(y)d\nu(y). \tag{4.7.4}$$

Thus g satisfies the same inequality (4.7.2) as quasi-nearly $\mathcal{H}$-subharmonic functions. In fact, the two concepts are the same (Exercise 4.8.5).

Quasi-nearly subharmonic functions, perhaps with a different terminology, have previously been considered by many authors. The concept itself dates back to C. Fefferman and E. M. Stein [21], who proved that if h is (Euclidean) harmonic, then $|h|^p$ is quasi-nearly subharmonic for all p, $0 < p < 1$. This result was also proved independently by Ü. Kuran [49]. For a non-negative subharmonic function f, J. Riihentaus [70] and N. Suzuki [92] independently proved that f^p is quasi-nearly subharmonic for all p, $0 < p < 1$.

We first prove that if $f \geq 0$ is a continuous quasi-nearly $\mathcal{H}$-subharmonic function, then f^p is quasi-nearly subharmonic for all p, $0 < p < \infty$. We will also prove an analogous result for $|\nabla^h H(x)|$ where H is $\mathcal{H}$-harmonic on $\mathbb{B}$. The proofs of both Theorems 4.7.3 and 4.7.4 are similar to the proofs given by M. Pavlović [62] for $\mathcal{M}$-harmonic functions in the unit ball of $\mathbb{C}^m$.

Theorem 4.7.3 *Let f be a continuous non-negative quasi-nearly $\mathcal{H}$-subharmonic function on $\mathbb{B}$. Then for all p, $0 < p < \infty$, there exists a constant $C(n,p,r)$, independent of f, such that*

$$f^p(a) \leq \frac{C(n,p,r)}{r^n} \int_{E(a,r)} f^p(x)d\tau(x)$$

for all $a \in \mathbb{B}$ and all r, $0 < r < 1$, where

$$C(n,p,r) = \begin{cases} c_n C^p (1 - r^2)^{-(n-1)(p-1)} & \text{if } p \geq 1, \\ C^{1/p} 2^{n/p} & \text{if } 0 < p < 1. \end{cases}$$

Proof. For $p > 1$, by Theorem 4.3.5 and Hölder's inequality,

$$f(a) \leq \frac{C}{r^n} \left(\int_{E(a,r)} f^p(x)d\tau(x) \right)^{1/p} \tau(B_r)^{1/q}.$$

But by inequality (3.3.8)

$$\tau(B_r) \le c_n \frac{r^n}{(1-r^2)^{n-1}}.$$

Therefore

$$f^p(a) \le \frac{1}{r^n} \frac{c_n C^p}{(1-r^2)^{(n-1)(p-1)}} \int_{E(a,r)} f^p(y) d\tau(y).$$

For the case $0 < p < 1$ we use a variation of the argument given in [62, Lemma 5]. Without loss of generality we take $C = 1$. Let $a \in \mathbb{B}$ and fix an r, $0 < r < 1$. Set

$$C_p^p = \int_{E(a,r)} f^p(x) d\tau(x),$$

and let $g(x) = f(x)/C_p$. Then $\int_{E(a,r)} g^p d\tau \le 1$ and

$$g(x) \le \frac{1}{s^n} \int_{E(x,s)} g(y) d\tau(y) \tag{4.7.5}$$

for all $x \in \mathbb{B}$ and all s, $0 < s < 1$. Define

$$A = \sup\{g^p(x)(r-\epsilon)^n : x \in E(a,\epsilon),\ 0 < \epsilon < r\}.$$

Since g is continuous, A is finite. Let $x \in E(a,\epsilon)$ and let $\epsilon < s < r$. Then by (4.7.5)

$$g(x)(s-\epsilon)^n \le \int_{E(x,s-\epsilon)} g(y) d\tau(y).$$

Since $d_{ph}(x,y) = |\varphi_x(y)|$ is a metric on $\mathbb{B}$ (Theorem 2.2.3),

$$E(x, s-\epsilon) \subset E(a,s) \subset E(a,r).$$

Therefore

$$\begin{aligned} g(x)(s-\epsilon)^n &\le \int_{E(a,s)} g^p(y) g^{1-p}(y) d\tau(y) \\ &\le \left[\frac{A}{(r-s)^n}\right]^{(1-p)/p} \int_{E(a,s)} g^p(y) d\tau(y) \\ &\le \left[\frac{A}{(r-s)^n}\right]^{(1-p)/p}. \end{aligned}$$

Take $s = (\epsilon + r)/2$. Then $s - \epsilon = r - s = (r-\epsilon)/2$, and thus

$$g^p(x)(r-\epsilon)^n \le 2^n A^{1-p}.$$

Taking the supremum over $x \in E(a,\epsilon)$, $0 < \epsilon < r$, gives

$$A \le 2^n A^{1-p} \qquad \text{or} \qquad A \le 2^{n/p}.$$

Therefore $g^p(a)(r-\epsilon)^n \le 2^{n/p}$ for all ϵ, $0 < \epsilon < r$. Thus

$$f^p(a) \le \frac{2^{n/p}}{r^n} \int_{E(a,r)} f^p(y)\,d\tau(y),$$

which is the desired inequality. □

Theorem 4.7.4 *Let H be $\mathcal{H}$-harmonic on $\mathbb{B}$. Then for all δ, $0 < \delta < \frac{1}{2}$, and all p, $0 < p < \infty$,*

(a) $|\nabla^h H(a)|^p \le \dfrac{C_{n,p}}{\delta^n} \displaystyle\int_{E(a,\delta)} |\nabla^h H(y)|^p d\tau(y)$, *and*

(b) $|\nabla^h H(a)|^p \le \dfrac{C_{n,p,\delta}}{\delta^n} \displaystyle\int_{E(a,\delta)} |H(y)|^p d\tau(y)$,

where $C_{n,p}$ and $C_{n,p,\delta}$ are constants depending only on p, n, and δ.

Proof. (a) Fix δ, $0 < \delta < \frac{1}{2}$, and let $0 < r < \delta$. Since $y \to H(\varphi_x(y))$ is $\mathcal{H}$-harmonic, we have

$$H(x) = H(\varphi_x(0)) = \int_{\mathbb{S}} H(\varphi_x(rt))\,d\sigma(t).$$

Hence

$$\frac{\partial H(x)}{\partial x_i} = \int_{\mathbb{S}} \frac{\partial H(\varphi_x(rt))}{\partial x_i} d\sigma(t).$$

Since $|y| \le \sum_i |y_i| \le n|y|$ for a vector $y = (y_1, \dots, y_n)$, we have

$$|\nabla H(x)| \le n \int_{\mathbb{S}} |\nabla_x H(\varphi_x(rt))|\,d\sigma(t),$$

and thus

$$|\nabla H(0)| \le n \int_{\mathbb{S}} |\nabla_x H(\varphi_x(rt))|_{x=0}\, d\sigma(t).$$

However, by Exercise 3.5.10

$$|\nabla_x H(\varphi_x(rt))|_{x=0} \le 2r^2 |\nabla H(-rt)| + (1+r^2)|\nabla H(-rt)| \le 4|\nabla H(-rt)|.$$

Therefore

$$|\nabla H(0)| \le 4n \int_{\mathbb{S}} |\nabla H(-rt)|\,d\sigma(t) = 4n \int_{\mathbb{S}} |\nabla H(rt)|\,d\sigma(t),$$

and by Hölder's inequality

$$|\nabla H(0)|^p \le C_n \int_{\mathbb{S}} |\nabla H(rt)|^p d\sigma(t)$$

for all $p \ge 1$. Multiplying by r^{n-1} and integrating from 0 to δ gives

$$|\nabla H(0)|^p \le \frac{C_n}{\delta^n} \int_{E(0,\delta)} |\nabla H(x)|^p d\nu(x).$$

But for $x \in E(0,\delta)$ with $\delta < \frac{1}{2}$ we have $\frac{3}{4} < (1 - |x|^2) < 1$. Therefore

$$|\nabla^h H(0)|^p \le \frac{C_n}{\delta^n} \int_{E(0,\delta)} |\nabla^h H(x)|^p d\tau(x),$$

where C_n is a constant depending only on n. The desired inequality for $p \ge 1$ is finally obtained by applying the above to $H \circ \varphi_a$. The result for $0 < p < 1$ now follows by Theorem 4.7.3.

(b) Let χ be a non-negative C^∞ radial function with $\operatorname{supp}\chi \subset B_{\delta/2}$ and $\int_{\mathbb{B}} \chi \, d\tau = 1$. Then by Theorem 4.5.4,

$$H(x) = (H * \chi)(x) = \int_{\mathbb{B}} H(y)\chi(\varphi_y(x))d\tau(y).$$

Thus

$$\nabla^h H(x) = \int_{\mathbb{B}} \nabla_x^h \chi(\varphi_y(x)) H(y) d\tau(y).$$

Hence,

$$|\nabla^h H(x)| \le C_\delta \sup_{y \in E(x,\delta/2)} |H(y)|,$$

where

$$C_\delta = \int_{E(x,\delta/2)} |\nabla_x^h \chi(\varphi_y(x))| d\tau(y),$$

which by the invariance of ∇^h and τ

$$= \int_{E(x,\delta/2)} |(\nabla^h h)(\varphi_x(y))| d\tau(y) = \int_{B_{\delta/2}} |\nabla^h \chi| d\tau.$$

Thus the constant C_δ is independent of x. Since $|H|$ is $\mathcal{H}$-subharmonic, by Theorem 4.7.3,

$$|H(y)|^p \le \frac{C_{n,p}}{\delta^n} \int_{E(y,\delta/2)} |H(y)|^p d\tau(y),$$

for all p, $0 < p < \infty$, and $C_{n,p}$ is a constant depending only on n, p. If $y \in E(x,\delta/2)$, then $E(y,\delta/2) \subset E(x,\delta)$. Therefore

$$|\nabla^h H(x)|^p \le \frac{C_{n,p,\delta}}{\delta^n} \int_{E(x,\delta)} |H(y)|^p d\tau(y).$$

□

4.8 Exercises

4.8.1. For $1 \le i,j \le n$ set

$$T_{i,j} = x_i \frac{\partial}{\partial x_j} - x_j \frac{\partial}{\partial x_i}.$$

(a) If f is $\mathcal{H}$-harmonic on $\mathbb{B}$, prove that $T_{i,j}f$ is $\mathcal{H}$-harmonic on $\mathbb{B}$.

(b) Prove that

$$|x|^2 |\nabla f(x)|^2 = |\langle x, \nabla f(x)\rangle|^2 + \sum_{i<j} |T_{i,j}f(x)|^2.$$

4.8.2. (a) For $\alpha \neq 0$ compute $\Delta_h(1-|x|^2)^\alpha$ and show that $(1-|x|^2)^\alpha$ is $\mathcal{H}$-superharmonic for $0 < \alpha \le (n-1)$, and $\mathcal{H}$-subharmonic for all $\alpha < 0$.

(b) Using (a), show that $\Delta_h(1-|x|^2)^{n-1} = -2n(n-1)(1-|x|^2)^n$.

4.8.3. (a) For $0 < \delta < \frac{1}{2}$, show that

$$E(x,\delta) \subset B(x, c_\delta(1-|x|^2)),$$

where $c_\delta = \delta/(1-\delta)$, and

(b)

$$B(x, \epsilon(1-|x|^2)) \subset E(x,\delta)$$

for all ϵ, $0 < \epsilon < \delta(1-\delta)$.

4.8.4. Show that $|\nabla^h f|$ is quasi-nearly $\mathcal{H}$-subharmonic if and only if $|\nabla f|$ is quasi-nearly $\mathcal{H}$-subharmonic.

4.8.5. Prove that a non-negative locally integrable function g is quasi-nearly $\mathcal{H}$-subharmonic if and only if g is quasi-nearly subharmonic.

4.8.6. If f is an eigenfunction of Δ_h with non-zero eigenvalue, that is, $\Delta_h f = \lambda f, \lambda \neq 0$, prove that for $0 < \delta < \frac{1}{2}$, $0 < p < \infty$,

$$|f(x)|^p \le \frac{C_{n,p,\delta}}{\delta^n} \int_{E(x,\delta)} |\nabla^h f(y)|^p \, d\tau(y),$$

where $C_{n,p,\delta}$ is a constant depending only on δ, p, and n.

Definition 4.8.1 *If μ is a regular Borel measure*[2] *on $\mathbb{B}$, the (invariant)* **Green potential** *of the measure μ, denoted G_μ, is defined on $\mathbb{B}$ by*

$$G_\mu(x) = \int_{\mathbb{B}} G_h(x,y)\, d\mu(y).$$

[2] A measure μ defined on the Borel subsets $\mathcal{B}$ of a locally compact Hausdorff space is called a **regular Borel measure** if (i) $\mu(K) < \infty$ for every compact set K, and for every Borel set B, (ii) $\mu(B) = \inf\{\mu(V) : B \subset V,\ V \text{ open}\}$, and (iii) $\mu(B) = \sup\{\mu(K) : K \subset B,\ K \text{ compact}\}$.

4.8.7. (a) If $G_\mu \not\equiv +\infty$, prove that the function G_μ is $\mathcal{H}$-superharmonic on $\mathbb{B}$ and $\mathcal{H}$-harmonic on $\mathbb{B} \setminus \operatorname{supp} \mu$.[3]

(b) If $\operatorname{supp} \mu$ is a compact subset of $\mathbb{B}$, prove that

$$\lim_{x \to t} G_\mu(x) = 0 \qquad \text{for every} \quad t \in \mathbb{S}.$$

4.8.8. Prove that $G_h(\cdot, x) \in L^p(\mathbb{B}, \nu)$ for all p, $0 < p < n/(n-1)$, and that

$$\int_{\mathbb{B}} G_h^p(y, x) d\nu(y) \leq A(n, p) G_h^p(0, x).$$

4.8.9. Let $f \in L^q(\mathbb{B}, \nu)$ for some $q > n$, and let

$$G_f(x) = \int_{\mathbb{B}} f(y) G_h(x, y) d\nu(y).$$

Prove that $\lim_{|x| \to 1} G_f(x) = 0$.

4.8.10. (*) (a) If f is continuous on $\overline{\mathbb{B}}$ and satisfies (4.3.3) for all $\psi \in \mathcal{M}(\mathbb{B})$, is f $\mathcal{H}$-harmonic on $\mathbb{B}$?

(b) If $f \in L^1(\mathbb{B})$ and f satisfies (4.3.3) for all $\psi \in \mathcal{M}(\mathbb{B})$, determine the values of the dimension $n > 2$, if any, for which f is $\mathcal{H}$-harmonic on $\mathbb{B} \subset \mathbb{R}^n$.

4.8.11. Let μ be a regular Borel measure on $\mathbb{B}$ satisfying

$$\int_{\mathbb{B}} (1 - |x|) d\mu(x) < \infty,$$

and set $V(x) = G_\mu^e(x)$, where G^e is the Euclidean Green's function on $\mathbb{B}$. In [8] D. H. Armitage proved that $V \in L^p(\mathbb{B})$ for all p, $0 < p < n/(n-1)$, with

$$\int_{\mathbb{B}} V^p(x) dx \leq A(n, p) V^p(0).$$

Investigate the analogous theorem for the invariant Green potential G_μ.

4.8.12. (*) Let V be as in the previous exercise. In [25] S. J. Gardiner proved that

(a) If $n \geq 2$ and $1 \leq p < (n-1)/(n-2)$, then

$$\lim_{r \to 1} (1 - r)^{(n-1)(p-1)} \int_{\mathbb{S}} V^p(rt) d\sigma(t) = 0.$$

(b) If $n \geq 3$ and $(n-1)/(n-2) \leq p < (n-1)/(n-3)$, then

$$\liminf_{r \to 1} (1 - r)^{(n-1)(p-1)} \int_{\mathbb{S}} V^p(rt) d\sigma(t) = 0.$$

[3] The **support** of μ, denoted $\operatorname{supp} \mu$, is the unique relatively closed subset F of $\mathbb{B}$ such that $\mu(\mathbb{B} \setminus F) = 0$ and $\mu(F \cap O) > 0$ for every open set O for which $O \cap F \neq \phi$.

The analogous result for Green potentials with regard to the Laplace–Beltrami operator on the unit ball in $\mathbb{C}^m$ was investigated by the author in [81] and [82]. Investigate this problem for hyperbolic Green potentials on $\mathbb{B}$. In this setting a suitable hypothesis on the measure μ is that it satisfies

$$\int_{\mathbb{B}} (1 - |x|^2)^{n-1} d\mu(x) < \infty.$$

4.8.13. (*) (a) Find necessary and sufficient conditions on an $\mathcal{H}$-harmonic function f such that $|\nabla^h f|^2$ is $\mathcal{H}$-subharmonic.

(b) If u is a Euclidean harmonic function on $\mathbb{B}$, then it is known that $\log|\nabla u|$ is subharmonic when $n = 2$, and that $|\nabla u|^p$ is subharmonic whenever $p \geq (n-2)/(n-1)$ when $n \geq 3$ [76, Theorem 4.14]. When $n \geq 3$, do there exist analogous results for $|\nabla^h f|$ where f is $\mathcal{H}$-harmonic on $\mathbb{B}$?

Exercises on the Upper Half-Space $\mathbb{H}$

4.8.14. Let $\Phi : \mathbb{H} \to \mathbb{B}$ be the mapping given by (2.3.2). If U is $\mathcal{H}$-harmonic on $\mathbb{B}$, prove that $V(x) = U(\Phi(x))$ is $\mathcal{H}$-harmonic on $\mathbb{H}$.

4.8.15. If U is a function on $\Omega \subset \mathbb{R}^n \setminus \{e_n\}$, define $\mathcal{K}(U)$ on $\Phi(\Omega)$ by

$$\mathcal{K}(U) = 2^{(n-2)/2}|x + e_n|^{2-n} U(\Phi(x)).$$

Prove that

(a) $\mathcal{K}[\mathcal{K}(u)] = u$.

(b) $\mathcal{K}[aU + bV] = a\mathcal{K}[U] + b\mathcal{K}[V]$ for $a, b \in \mathbb{R}$.

(c) If U is a (Euclidean) harmonic function on Ω, then $\mathcal{K}[U]$ is a (Euclidean) harmonic function on $\Phi(\Omega)$.

5

The Poisson Kernel and Poisson Integrals

We begin the chapter by using Green's formula to derive the hyperbolic Poisson kernel P_h on $\mathbb{B}$. In Section 5.2 we include a theorem of P. Jaming that expresses the Euclidean Poisson kernel as an integral involving the hyperbolic Poisson kernel. In Section 5.3 we solve the Dirichlet problem for $\mathbb{B}$, and in Section 5.4 we use a method due to W. Rudin to solve the Dirichlet problem for B_r. As in [72] we characterize the eigenfunctions of Δ_h in Section 5.5, and in Section 5.6 we derive the Poisson kernel for the upper half-space $\mathbb{H}$.

5.1 The Poisson Kernel for Δ_h

In this section we provide a heuristic argument for the formula for the Poisson kernel on $\mathbb{B}$. We begin by stating **Green's formula** (see [74]) for the invariant Laplacian: if Ω is an open subset of $\mathbb{B}$, $\overline{\Omega} \subset \mathbb{B}$, whose boundary is sufficiently smooth, then if $u, v \in C^2(\Omega) \cap C^1(\overline{\Omega})$,

$$\int_{\Omega} (u\Delta_h v - v\Delta_h u)d\tau = \int_{\partial\Omega} (uD_{\tilde{n}}v - vD_{\tilde{n}}u)\, d\tilde{\sigma}, \qquad (5.1.1)$$

where $\tilde{\sigma}$ is the surface element on $\partial\Omega$ with respect to the hyperbolic metric, and $D_{\tilde{n}}$ denotes the **normal derivative** in the outward normal direction again with respect to the hyperbolic metric. With $\Omega = B_r$, the above becomes

$$\int_{B_r} (u\Delta_h v - v\Delta_h u)d\tau = \int_{S_r} (uD_{\tilde{n}}v - vD_{\tilde{n}}u)\, d\tilde{\sigma}. \qquad (5.1.2)$$

For the surface S_r given by $f(x) = 0$, where $f(x) = |x|^2 - r^2$, the normal derivative $D_{\tilde{n}}$ of v is given by

$$D_{\tilde{n}}v(r\zeta) = \frac{\langle \nabla^h v(r\zeta), \nabla^h f(r\zeta)\rangle}{|\nabla^h f(r\zeta)|} = (1 - r^2)\langle \nabla v(r\zeta), \zeta\rangle.$$

Also, by Theorem 4.1.1 (a),

$$\begin{aligned}\int_{B_r} \Delta_h u(x)d\tau(x) &= nr^{n-1}(1-r^2)^{2-n}\frac{d}{dr}\int_{\mathbb{S}} f(r\zeta)\,d\sigma(\zeta)\\ &= nr^{n-1}(1-r^2)^{2-n}\int_{\mathbb{S}}\langle\nabla u(r\zeta),\zeta\rangle\,d\sigma(\zeta)\\ &= nr^{n-1}(1-r^2)^{1-n}\int_{\mathbb{S}} D_{\tilde{n}}u(r\zeta)\,d\sigma(\zeta).\end{aligned}$$

Thus the surface element $d\tilde{\sigma}$ on S_r is given by

$$d\tilde{\sigma}(r\zeta) = \frac{nr^{n-1}}{(1-r^2)^{n-1}}\,d\sigma(\zeta). \tag{5.1.3}$$

Suppose now that u is $\mathcal{H}$-harmonic on $\mathbb{B}$, and for the sake of simplicity C^1 on $\overline{\mathbb{B}}$. Let $a \in \mathbb{B}$ be arbitrary, and let $r_o > 0$ be such that $a \in B_r$ for all r, $r_o \le r < 1$. Choose $\epsilon > 0$ such that $E(a,\epsilon) \subset B_{r_o}$. Then by Green's formula applied to $\Omega_\epsilon = B_r \setminus E(a,\epsilon)$ with $v(x) = g_h(\varphi_a(x))$ where g_h is given by (3.2.1),

$$\int_{S_r}\{uD_{\tilde{n}}v - vD_{\tilde{n}}u\}\,d\tilde{\sigma} = \int_{\varphi_a(S_\epsilon)}\{uD_{\tilde{n}}v - vD_{\tilde{n}}u\}d\tilde{\sigma},$$

which by the $\mathcal{M}$-invariance of $\tilde{\sigma}$

$$\begin{aligned}&= \int_{S_\epsilon}\{(u\circ\varphi_a)D_{\tilde{n}}(v\circ\varphi_a) - (v\circ\varphi_a)D_{\tilde{n}}(u\circ\varphi_a)\}\,d\tilde{\sigma}\\ &= \int_{S_\epsilon}\{(u\circ\varphi_a)D_{\tilde{n}}g_h - g_h D_{\tilde{n}}(u\circ\varphi_a)\}d\tilde{\sigma}.\end{aligned}$$

Clearly

$$\lim_{\epsilon\to 0}\int_{S_\epsilon} g_h D_{\tilde{n}}(u\circ\varphi_a)d\tilde{\sigma} = 0.$$

On the other hand, since g_h is radial,

$$D_{\tilde{n}}g_h(\epsilon\zeta) = (1-\epsilon^2)g_h'(\epsilon) = -\frac{(1-\epsilon^2)^{n-1}}{n\epsilon^{n-1}}.$$

Thus by (5.1.3)

$$\begin{aligned}\lim_{\epsilon\to 0}\int_{S_\epsilon}(u\circ\varphi_a)D_{\tilde{n}}g_h\,d\tilde{\sigma} &= -\lim_{\epsilon\to 0}\int_{\mathbb{S}} u(\varphi_a(\epsilon\zeta))\,d\sigma(\zeta)\\ &= -u(a).\end{aligned}$$

Hence for any $r > r_o$,

$$u(a) = -\int_{S_r} [uD_{\tilde{n}}v - vD_{\tilde{n}}u]\, d\tilde{\sigma}.$$

Since $G_h(a,r) \approx (1-r^2)^{n-1}$ and $u \in C^1(\overline{\mathbb{B}})$,

$$\lim_{r\to 1}\int_{S_r} vD_{\tilde{n}}u\, d\tilde{\sigma} \approx \lim_{r\to 1}(1-r^2)\int_{\mathbb{S}}\langle \nabla u(r\zeta),\zeta\rangle\, d\sigma(\zeta) = 0.$$

Thus setting $G_a(x) = G_h(a,x)$,

$$\begin{aligned} u(a) &= -\lim_{r\to 1}\int_{S_r} uD_{\tilde{n}}G_a d\tilde{\sigma} \\ &= -\lim_{r\to 1} nr^{n-1}(1-r^2)^{2-n}\int_{\mathbb{S}} u(r\zeta)\langle \nabla G_a(r\zeta),\zeta\rangle\, d\sigma(\zeta) \\ &= \int_{\mathbb{S}} P_h(a,\zeta)u(\zeta)\, d\sigma(\zeta), \end{aligned}$$

where

$$P_h(a,\zeta) = -\lim_{r\to 1} nr^{n-1}(1-r^2)^{2-n}\langle \nabla G_a(r\zeta),\zeta\rangle. \tag{5.1.4}$$

Our next step is to compute the above limit. Since $G_a(x) = g_h(|\varphi_a(x)|)$, where g is the radial function defined by (3.2.1),

$$\begin{aligned} \nabla G_a(x) &= -\frac{(1-|\varphi_a(x)|^2)^{n-2}}{n|\varphi_a(x)|^{n-1}}\nabla|\varphi_a(x)| \\ &= -\frac{(1-|a|^2)^{n-2}(1-|x|^2)^{n-2}}{n|x-a|^{n-1}\rho(x,a)^{\frac{1}{2}(n-3)}}\nabla|\varphi_a(x)|. \end{aligned}$$

Using identity (2.1.7) we have

$$2|\varphi_a(x)|\nabla|\varphi_a(x)| = \nabla\left[\frac{|x-a|^2}{\rho(x,a)}\right].$$

From this it now follows that

$$\nabla|\varphi_a(x)| = \frac{(x-a)(1-|x|^2)(1-|a|^2) + x|x-a|^2(1-|a|^2)}{|x-a|\rho(x,a)^{\frac{3}{2}}}.$$

Therefore

$$\begin{aligned} &-nr^{n-1}(1-r^2)^{2-n}\langle \nabla G_a(r\zeta),\zeta\rangle \\ &\quad= \frac{r^{n-1}(1-|a|^2)^{n-2}}{|r\zeta-a|^n\rho(r\zeta,a)^{\frac{n}{2}}}\Big[(1-r^2)(1-|a|^2)\langle r\zeta-a,\zeta\rangle + r|r\zeta-a|^2(1-|a|^2)\Big]. \end{aligned}$$

Thus

$$-\lim_{r\to 1} r^{n-1}(1-r^2)^{2-n}\langle \nabla G_a(r\zeta),\zeta\rangle = \frac{(1-|a|^2)^{n-1}|\zeta-a|^2}{|\zeta-a|^n\rho(\zeta,a)^{\frac{n}{2}}},$$

which since $\rho(\zeta,a)=|\zeta-a|^2$ gives

$$P_h(a,\zeta)=\left(\frac{1-|a|^2}{|\zeta-a|^2}\right)^{n-1}, \qquad (a,\zeta)\in\mathbb{B}\times\mathbb{S}. \tag{5.1.5}$$

Definition 5.1.1 *The function P_h on $\mathbb{B}\times\mathbb{S}$ defined by (5.1.5) is called the* **invariant Poisson kernel** *for Δ_h on $\mathbb{B}$.*

A tedious, but straightforward computation proves that for each $t\in\mathbb{S}$, the function $x\mapsto P_h(x,t)$ is $\mathcal{H}$-harmonic on $\mathbb{B}$. This is outlined in greater detail in Section 5.5.

In contrast to the above, the Poisson kernel P_e for the ordinary Laplacian Δ on $\mathbb{B}\times\mathbb{S}$ is given by

$$P_e(x,t)=\frac{1-|x|^2}{|x-t|^n}, \qquad (x,t)\in\mathbb{B}\times\mathbb{S},$$

whereas the Poisson kernel $\widetilde{\mathcal{P}}$ for the invariant Laplacian $\widetilde{\Delta}$ on the Hermitian ball $\mathbb{B}$ in $\mathbb{C}^n$ is given by

$$\widetilde{\mathcal{P}}(z,t)=\frac{(1-|z|^2)^n}{|1-\langle z,t\rangle|^{2n}}, \qquad (z,t)\in\mathbb{B}\times\mathbb{S}.$$

It is only in the unit disc $\mathbb{D}$ in $\mathbb{R}^2$ that all three agree.

5.2 Relationship between the Euclidean and Hyperbolic Poisson Kernel

Writing the hyperbolic Poisson kernel P_h as

$$P_h(x,t)=\frac{(1-|x|^2)}{|x-t|^n}\frac{(1-|x|^2)^{n-2}}{|x-t|^{n-2}},$$

we obtain that

$$(1-|x|)^{n-2}P_e(x,t)\le P_h(x,t)\le 2^{n-2}P_e(x,t).$$

In this section we prove the following much stronger result due to P. Jaming [42].

Theorem 5.2.1 *There exists a function* $\eta : [0,1] \times [0,1] \to \mathbb{R}^+$ *such that*

(a) $P_e(r\zeta, t) = \int_0^1 \eta(r,\rho) P_h(\rho r\zeta, t) d\rho,$ *and*

(b) *there exists a constant C, independent of r, such that*

$$\int_0^1 \eta(r,\rho) d\rho \le C$$

for every $r \in [0,1]$.

Proof. Since

$$\int_0^\infty \frac{t^{n/2-2}}{(1+t)^{n-1}} = \frac{\Gamma(\frac{n}{2}-1)\Gamma(\frac{n}{2})}{\Gamma(n-1)},$$

where Γ is the gamma function, we have

$$\frac{1}{(x+y)^{\frac{n}{2}}} = c_n \int_0^\infty \frac{s^{n/2-2}}{(x+y+s)^{n-1}} ds.$$

Since

$$P_e(r\zeta, t) = \frac{1-r^2}{[(1-r)^2 + 2r(1-\langle\zeta, t\rangle)]^{\frac{n}{2}}},$$

writing $X = 2(1-\langle\zeta, t\rangle)$ we obtain

$$\begin{aligned} P_e(r\zeta, t) &= \frac{(1-r^2)}{|(1-r)^2 + rX]^{\frac{n}{2}}} = \frac{(1-r^2)}{r^{\frac{n}{2}}[(1-r)^2/r + X]^{\frac{n}{2}}} \\ &= c_n \frac{(1-r^2)}{r^{\frac{n}{2}}} \int_0^\infty \frac{s^{n/2-2} ds}{[X + (1-r)^2/r + s]^{n-1}}. \end{aligned}$$

The change of variable

$$s = \frac{(1-\rho)^2}{\rho} - \frac{(1-r)^2}{r} = \frac{(r-\rho)(1-\rho r)}{\rho r}$$

gives

$$\begin{aligned} P_e(r\zeta, t) &= c_n \frac{(1-r^2)}{r^{\frac{n}{2}}} \int_0^r \frac{[(r-\rho)(1-\rho r)]^{n/2-2}}{\left[X + (1-\rho)^2/\rho\right]^{n-1} (\rho r)^{n/2-2}} \frac{(1-\rho^2)}{\rho^2} d\rho \\ &= c_n \frac{(1-r^2)}{r^{n-2}} \int_0^r \frac{[(r-\rho)(1-\rho r)]^{n/2-2} \rho^{n/2-1}(1-\rho^2)}{[\rho X + (1-\rho)^2]^{n-1}} d\rho \\ &= c_n \frac{(1-r)^2}{r^{n-2}} \int_0^r P_h(\rho\zeta, t) \frac{[(r-\rho)(1-\rho r)]^{n/2-2} \rho^{n/2-1}}{(1-\rho^2)^{n-2}} d\rho. \end{aligned}$$

Finally with the change of variable $\rho = rs$, $s \in (0,1)$, we obtain

$$P_e(r\zeta, t) = \int_0^1 P_h(rs\zeta, t)\, \eta(r,s)\, ds,$$

where

$$\eta(r,s) = c_n(1-r^2)\frac{[(1-s)(1-sr^2)]^{n/2-2}s^{n/2-1}}{(1-r^2s^2)^{n-2}}. \tag{5.2.1}$$

For the proof of (b) we first note that

$$(1-r^2s^2)^{n-2} \geq (1-s)^{n/2-2}(1-sr^2)^{n/2}.$$

Therefore,

$$I(r) = \int_0^1 \eta(r,s)ds \leq c_n(1-r^2)\int_0^1 (1-sr^2)^{-2}s^{n/2-1}ds,$$

which by the change of variable $u = sr^2$

$$= c_n\frac{(1-r^2)}{r^n}\int_0^{r^2}(1-u)^{-2}u^{n/2-1}du.$$

For $r^2 < \frac{1}{2}$,

$$I(r) \leq C_n\frac{(1-r^2)}{r^n}\int_0^{r^2} u^{n/2-1}du \leq \frac{2}{n}C_n(1-r^2) \leq C_n.$$

On the other hand, for $r^2 > \frac{1}{2}$,

$$I(r) \leq C_n(1-r^2)\int_0^{r^2}(1-u)^{-2}du \leq C_n.$$

Thus $I(r) \leq C$ for all $r \in [0,1]$. □

Remark 5.2.2 *One other interesting relationship between $\mathcal{H}$-harmonic and Euclidean harmonic functions is the following result of H. Samii* [73]*: if u is $\mathcal{H}$-harmonic on $\mathbb{B}$, then there exists a Euclidean harmonic function v with $v(0) = 0$ such that*

$$u(x) = u(0) + \int_0^1 v(tx)[(1-t)(1-tr^2)]^{n/2-1}\frac{dt}{t} \tag{5.2.2}$$

for all $x \in \mathbb{B}$.

5.3 The Dirichlet Problem for $\mathbb{B}$

We summarize some of the properties of the invariant Poisson kernel in the following lemma. These are analogous to the properties of the Poisson kernel for Δ on $\mathbb{B}$.

Lemma 5.3.1 *The Poisson kernel P_h on $\mathbb{B} \times \mathbb{S}$ satisfies the following:*

(a) *For fixed $t \in \mathbb{S}$, $x \mapsto P_h(x,t)$ is $\mathcal{H}$-harmonic on $\mathbb{B}$,*

(b) *$P_h(r\zeta, t) = P_h(rt, \zeta)$ for all $t, \zeta \in \mathbb{S}$,*

(c) $\displaystyle\int_{\mathbb{S}} P_h(x,t)\, d\sigma(t) = 1$, *and*

(d) *For fixed $\zeta \in \mathbb{S}$ and $\delta > 0$,* $\displaystyle\lim_{\substack{x\to\zeta \\ x\in\mathbb{B}}} \int_{|t-\zeta|>\delta} P_h(x,t)\, d\sigma(t) = 0.$

Proof. As indicated above, (a) follows by computation, and (b) is almost obvious. Writing $x = r\zeta$, (c) follows by (b) and the mean-value property for $\mathcal{H}$-harmonic functions. The proof of (d) is again standard. □

Definition 5.3.2 *For $f \in L^1(\mathbb{S})$, the* **Poisson integral** *of f denoted $P_h[f]$ is defined by*

$$P_h[f](x) = \int_{\mathbb{S}} P_h(x,t) f(t)\, d\sigma(t).$$

Similarly, if μ is a finite signed Borel[1] *measure on $\mathbb{S}$, the Poisson integral of μ will be denoted by $P_h[\mu]$, that is,*

$$P_h[\mu](x) = \int_{S} P_h(x,t)\, d\mu(t).$$

Since the function $t \mapsto P_h(x,t)$ is continuous on $\mathbb{S}$, the above integrals exist and are finite for all $x \in \mathbb{B}$. Furthermore, as a consequence of the mean-value property, the function $P_h[\mu](x)$ is $\mathcal{H}$-harmonic on $\mathbb{B}$. Since the Poisson kernel is an approximate identity, we have the following theorem.

Theorem 5.3.3 *Let f be a bounded measurable function on $\mathbb{S}$. Then*

$$\lim_{\substack{x\to\zeta \\ x\in\mathbb{B}}} P_h[f](x) = f(\zeta)$$

at each $\zeta \in \mathbb{S}$ where f is continuous.

Proof. Suppose $|f(t)| \le M$ for all $t \in \mathbb{S}$ and that f is continuous at $\zeta \in \mathbb{S}$. Then given $\epsilon > 0$ there exists a $\delta > 0$ such that $|f(t) - f(\zeta)| < \epsilon$ for all $t \in \mathbb{S}$ with $|t - \zeta| < \delta$. Set $Q(\zeta, \delta) = \{t \in \mathbb{S} : |t - \zeta| < \delta\}$. Then

$$|P_h[f](x) - f(\zeta)| = \left| \int_{\mathbb{S}} P_h(x,t) f(t) d\sigma(t) - f(\zeta) \right|,$$

which by Lemma 5.3.1(c),

[1] Recall that the **Borel subsets** of S are the smallest σ-algebra of subsets of S such that every continuous function is measurable.

$$\leq \int_{\mathbb{S}} P_h(x,t)|f(t)-f(\zeta)|\,d\sigma(\zeta)$$
$$\leq \int\limits_{Q(\zeta,\delta)} P_h(x,t)|f(t)-f(\zeta)|d\sigma(t) + 2M \int\limits_{\mathbb{S}\setminus Q(\zeta,\delta)} P_h(x,t)d\sigma(t)$$
$$\leq \epsilon + 2M \int\limits_{|t-\zeta|\geq\delta} P_h(x,t)d\sigma(t).$$

The result now follows by Lemma 5.3.1(d). □

Corollary 5.3.4 **(a)** *If $f \in C(\mathbb{S})$, then the function F defined by*

$$F(x) = \begin{cases} P_h[f](x), & x \in \mathbb{B}, \\ f(x), & x \in \mathbb{S}, \end{cases}$$

is $\mathcal{H}$-harmonic on $\mathbb{B}$ and continuous on $\overline{\mathbb{B}}$ with $\| F\|_\infty = \|f\|_\infty$.[2]

(b) *Conversely, if f is $\mathcal{H}$-harmonic on $\mathbb{B}$ and continuous on $\overline{\mathbb{B}}$, then $f(x) = P_h[f](x)$.*

Proof. Part (a) is an immediate consequence of the previous theorem, and the proof of (b) is a consequence of the maximum principle (Theorem 4.1.6) applied to $F(x) = f(x) - P_h[f](x)$. □

An immediate consequence of the previous is the following analogue of [72, Theorem 3.3.8].

Theorem 5.3.5 *If $f \in L^1(\mathbb{S})$, then $P_h[f\circ\psi] = P_h[f]\circ\psi$ for all $\psi \in \mathcal{M}(\mathbb{B})$.*

Proof. Since $C(\mathbb{S})$ is dense in $L^1(\mathbb{S})$, it suffices to prove the result for continuous functions on $\mathbb{S}$. If $f \in C(\mathbb{S})$, then by the previous theorem $F(x) = P_h[f](x)$ is $\mathcal{H}$-harmonic on $\mathbb{B}$ and continuous on $\overline{\mathbb{B}}$. Suppose $\psi \in \mathcal{M}(\mathbb{B})$. Since ψ is continuous on $\overline{\mathbb{B}}$, $F\circ\psi$ is also $\mathcal{H}$-harmonic on $\mathbb{B}$, continuous on $\overline{\mathbb{B}}$ with $\lim_{x\to\zeta}(F\circ\psi)(x) = f(\psi(\zeta))$ for all $\zeta \in \mathbb{S}$. Thus $(F\circ\psi)(x) = P_h[f\circ\psi](x)$. On the other hand, $(F\circ\psi)(x) = F(\psi(x)) = P_h[f](\psi(x))$, which proves the result. □

If in the above proof we take $\psi = \varphi_a$ and $x = 0$, then

$$\int_S f(\varphi_a(t))\,d\sigma(t) = \int_{\mathbb{S}} P_h(a,t)f(t)\,d\sigma(t) \tag{5.3.1}$$

for all $f \in L^1(\mathbb{S})$.

[2] See (5.4.1) for the definition of $\|\ \|_\infty$.

Theorem 5.3.6 *Let ν be a finite signed Borel measure on $\mathbb{S}$. Then*
(a) *for all $\psi \in C(\mathbb{S})$,*

$$\lim_{r\to 1}\int_{\mathbb{S}} P_h[\nu](r\zeta)\psi(\zeta)d\sigma(\zeta) = \int_S \psi(\zeta)d\nu(\zeta).$$

(b) *If $P_h[\nu](x) = 0$ for all $x \in \mathbb{B}$, then $\nu = 0$.*

Proof. (a) Since $P_h(rt,\zeta) = P_h(r\zeta, t)$, by Fubini's theorem we have

$$\lim_{r\to 1}\int_{\mathbb{S}} P_h[\nu](rt)\psi(t)d\sigma(t) = \lim_{r\to 1}\int_{\mathbb{S}} P_h[\psi](r\zeta)\,d\nu(\zeta),$$

which by Corollary 5.3.4

$$= \int_{\mathbb{S}} \psi(\zeta)\,d\nu(\zeta).$$

(b) By (a), the hypothesis implies that $\int_{\mathbb{S}} \psi d\nu = 0$ for every continuous function ψ on $\mathbb{S}$. Suppose $\nu \neq 0$. Let $(P, \widetilde{P})$ be a Hahn decomposition for ν and let ν^+ and ν^- be the positive and negative variation of ν, that is,

$$\nu^+(E) = \nu(E\cap P) \quad \text{and} \quad \nu^-(E) = -\nu(E\cap \widetilde{P})$$

for all measurable sets E. (See [71] for details.) Since $\nu \neq 0$ we can assume, without loss of generality, that $\nu^+(P) > 0$. Choose F a compact subset of P and $\epsilon > 0$ such that $\nu^+(F) - \epsilon > 0$. Since all Borel measures on $\mathbb{S}$ are regular, there exists a compact subset C of $\widetilde{P}$ such that $\nu^-(\widetilde{P}\setminus C) < \epsilon$. Let $U = \widetilde{C}$. Then $F \subset U$, and by Urysohn's lemma (see [71, Chapter 3]) there exists a continuous function ψ on $\mathbb{S}$ with $0 \le \psi \le 1$, $\psi = 1$ on F, and $\operatorname{supp}\psi \subset U$. Therefore

$$\nu^+(F) = \int_F \psi\,d\nu^+ \le \int_P \psi\,d\nu$$

and

$$\int_{\widetilde{P}} \psi\,d\nu = \int_{\widetilde{P}\cap U} \psi\,d\nu^- \le \nu^-(\widetilde{P}\setminus C) < \epsilon.$$

Therefore,

$$0 < \nu^+(F) - \epsilon \le \int_P \psi\,d\nu - \int_{\widetilde{P}} \psi\,d\nu = \int_{\mathbb{S}} \psi\,d\nu = 0,$$

which is a contradiction. □

There is a significant difference between invariant Poisson integrals and solutions of the classical Dirichlet problem for the Laplacian Δ. Since Δ is uniformly elliptic, if $f \in C^\infty(\mathbb{S})$ and F is the Euclidean Poisson integral of f, then $F \in C^\infty(\overline{\mathbb{B}})$. The following example shows that this fails dramatically for invariant Poisson integrals.

Example 5.3.7 To illustrate the above we compute the invariant Poisson integral of the function $f(t) = t_1^2$ for $n = 3$. In this case

$$F(x) = (1 - |x|^2)^2 \int_{\mathbb{S}} \frac{t_1^2}{|x - t|^4} \, d\sigma(t).$$

With $x = re_1$, where $e_1 = (1, 0, 0)$,

$$F(re_1) = (1 - r^2)^2 \int_{\mathbb{S}} \frac{t_1^2}{(1 + r^2 - 2rt_1)^2} \, d\sigma(t),$$

which since the integrand is a function of t_1 only (see [10, p. 216])

$$\begin{aligned} &= \tfrac{1}{2}(1 - r^2)^2 \int_{-1}^{1} \frac{x^2}{(1 + r^2 - 2rx)^2} \, dx \\ &= \frac{1}{4r^3} \left[2r(1 + r^4) + (1 + r^2)(1 - r^2)^2 \log\left(\frac{1 - r}{1 + r}\right) \right]. \end{aligned}$$

Note, at $r = 0$, the term in brackets is of the form $\frac{4}{3}r^3 + O(r^5)$, and thus $F(re_1)$ is indeed continuous at 0. Even though $f(t) = t_1^2$ is C^∞ on $\mathbb{S}$, the function $F(re_1)$ is not C^2 at the boundary point e_1. A formula valid for all $x \in \mathbb{B}$ will be given in Example 6.1.3.

5.4 The Dirichlet Problem for B_r

In Theorem 5.3.4 we proved that the Dirichlet problem for Δ_h was solvable for the unit ball $\mathbb{B}$. In the following theorem we prove that the Dirichlet problem is also solvable for B_r, where for $0 < r < 1$,

$$B_r = \{y : |y| < r\} = \{rx : x \in \mathbb{B}\}.$$

Similarly, set

$$S_r = \{y : |y| = r\} = \{rt : t \in \mathbb{S}\}.$$

Our method of proof is analogous to that used by W. Rudin in [72] in proving the same theorem for $\mathcal{M}$-harmonic functions in the unit ball of $\mathbb{C}^n$.

Theorem 5.4.1 *Fix r, $0 < r < 1$. If $f \in C(S_r)$, then there exists $F \in C(\overline{B_r})$ such that*

(a) $\Delta_h F = 0$ *in* B_r,

(b) $F(rt) = f(rt)$ *for all* $t \in \mathbb{S}$, *and*

(c) $F(0) = \displaystyle\int_{\mathbb{S}} f(rt) d\sigma(t)$.

Proof. Let $\mathcal{H}(\overline{B_r})$ denote the class of functions F which are $\mathcal{H}$-harmonic in B_r and continuous on $\overline{B_r}$. For $F \in \mathcal{H}(\overline{B_r})$ set

$$\|F\|_\infty = \sup\{|F(y)| : y \in \overline{B_r}\} = \sup\{|F(rt)| : t \in \mathbb{S}\}. \tag{5.4.1}$$

The last equality follows from the maximum principle for $\mathcal{H}$-subharmonic functions. Let

$$\mathcal{H}(S_r) = \{f \in C(S_r) : f(rt) = F(rt) \quad \text{for some} \quad F \in \mathcal{H}(\overline{B_r})\}.$$

Then $\mathcal{H}(S_r)$ with the sup norm $\|\cdot\|_\infty$ is a closed subspace of $C(S_r)$, which by (5.4.1) is isomorphic to $\mathcal{H}(\overline{B_r})$.

We now show that $\mathcal{H}(S_r) = C(S_r)$. If not, then by the Hahn–Banach theorem there exists a non-trivial continuous linear functional γ on $C(S_r)$ such that $\gamma(f) = 0$ for all $f \in \mathcal{H}(S_r)$. Thus by the Riesz representation theorem there exists a finite signed Borel measure ν on $\mathbb{S}$ such that

$$\int_{\mathbb{S}} f(rt)\, d\nu(t) = 0 \qquad \text{for all} \quad f \in \mathcal{H}(S_r).$$

In particular,

$$\int_{\mathbb{S}} P_h[\psi](rt)\, d\nu(t) = 0 \qquad \text{for all} \quad \psi \in C(\mathbb{S}).$$

Thus by Fubini's theorem

$$\int_{\mathbb{S}} P_h[\psi](rt)\, d\nu(t) = \int_{\mathbb{S}} P_h[\nu](r\zeta)\psi(\zeta)d\sigma(\zeta) = 0$$

for all $\psi \in C(\mathbb{S})$. Therefore $P_h[\nu](r\zeta) = 0$ for all $\zeta \in \mathbb{S}$. Hence by the maximum principle $P_h[\nu](x) = 0$ for all $x \in B_r$. Since $\mathcal{H}$-harmonic functions are real analytic,[3] $P_h[\nu](x) = 0$ for all $x \in \mathbb{B}$. Thus by Theorem 5.3.6 we have that $\nu = 0$.

Finally, to show that $F(0) = \int_{\mathbb{S}} f(rt) d\sigma(t)$, define γ_0 on $\mathcal{H}(\overline{B_r})$ by $\gamma_0(F) = F(0)$. This defines a continuous linear functional on $C(S_r)$ in the obvious way. Thus there exists a signed Borel measure ν on $\mathbb{D}$ such that

$$\int_{\mathbb{S}} F(rt) d\nu(t) = F(0) \tag{5.4.2}$$

for all $F \in \mathcal{H}(\overline{B_r})$. Let $\mu = \nu - \sigma$. Then if $\psi \in C(\mathbb{S})$,

$$\int_{\mathbb{S}} P_h[\mu](rt)\psi(t)d\sigma(t) = \int_{\mathbb{S}} \left[\int_{\mathbb{S}} P_h(rt,\zeta)d\nu(\zeta) - P_h(0,\zeta)\right] \psi(t)d\sigma(t) = 0.$$

[3] The fact that $\mathcal{H}$-harmonic functions are real analytic follows from the following theorem of Hörmander [40, Theorem 7.5.1]: if L is an elliptic differential operator with real analytic coefficients, then every solution of $Lu = 0$ is real analytic.

The last equality follows by (5.4.2) since $P_h(\cdot, t) \in \mathcal{H}(\overline{B_r})$. Thus, as above, $\mu = 0$, that is, $\nu = \sigma$. □

In Remark 4.1.4 it was pointed out that if $f \in C^2(\mathbb{B})$ is $\mathcal{H}$-subharmonic on $\mathbb{B}$, then the integral mean

$$M(f,r) = \int_{\mathbb{S}} f(rt)d\sigma(t)$$

is a non-decreasing function of r, $0 < r < 1$. We now use the previous theorem to prove that this in fact holds for all $\mathcal{H}$-subharmonic functions.

Theorem 5.4.2 *If f is $\mathcal{H}$-subharmonic on $\mathbb{B}$, then for every $a \in \mathbb{B}$,*

$$\int_{\mathbb{S}} f(\varphi_a(rt))d\sigma(t)$$

is a non-decreasing function of r, $0 < r < 1$.

Proof. Without loss of generality we take $a = 0$. Suppose $0 < r_1 < r_2 < 1$. We first suppose that f is continuous on $\mathbb{B}$. By Theorem 5.4.1 there exists a function G which is $\mathcal{H}$-harmonic on B_{r_2} with $G(r_2t) = f(r_2t)$. Thus by the maximum principle $f(x) \le G(x)$ for all $x \in B_{r_2}$. Therefore

$$\int_{\mathbb{S}} f(r_1t)d\sigma(t) \le \int_{\mathbb{S}} G(r_1t)d\sigma(t) = G(0) = \int_{\mathbb{S}} G(r_2t)d\sigma(t) = \int_{\mathbb{S}} f(r_2t)d\sigma(t).$$

For arbitrary f, since f is upper semicontinuous, by Theorem 4.3.2(d) there exists a decreasing sequence $\{f_n\}$ of continuous functions on S_{r_2} with $f_n(r_2t) \to f(r_2t)$ for all $t \in \mathbb{S}$. For each n let $F_n \in \mathcal{H}(\overline{B_{r_2}})$ be such that $F_n(r_2t) = f_n(r_2t)$ for all $t \in \mathbb{S}$. Since $F_n \ge f$ on S_{r_2}, by the maximum principle $F_n \ge f$ on B_{r_2}. Therefore,

$$\int_{\mathbb{S}} f(r_1t)d\sigma(t) \le \int_{\mathbb{S}} F_n(r_2t)d\sigma(t) = \int_{\mathbb{S}} f_n(r_2t)d\sigma(t).$$

The result now follows by letting $n \to \infty$. □

5.5 Eigenfunctions of Δ_h

In this section we consider the eigenfunctions of the invariant Laplacian Δ_h. In the setting of rank one noncompact symmetric spaces this topic has previously been considered by K. Minemura [58]. Our approach follows that of W. Rudin in [72].

For fixed $t \in \mathbb{S}$ and $\alpha \in \mathbb{R}$, consider the function $P^\alpha(x,t)$. Since $x \to P_h(x,t)$ is $\mathcal{H}$-harmonic, by Exercise 3.5.3(b)

$$\Delta_h P_h^\alpha(x,t) = \alpha(\alpha-1)P_h(x,t)^{\alpha-2}|\nabla_x^h P_h(x,t)|^2.$$

But $|\nabla_x^h P_h(x,t)|^2 = (1-|x|^2)^2|\nabla_x P_h(x,t)|^2$. It is left as an exercise (Exercise 5.7.3) to show that

$$|\nabla_x P_h(x,t)|^2 = 4(n-1)^2\frac{(1-|x|^2)^{2n-4}}{|x-t|^{4n-4}}. \tag{5.5.1}$$

Therefore

$$\begin{aligned}|\nabla_x^h P_h(x,t)|^2 &= 4(n-1)^2\frac{(1-|x|^2)^{2n-2}}{|x-t|^{4n-4}}\\ &= 4(n-1)^2 P_h(x,t)^2\end{aligned}$$

and

$$\Delta_h P_h^\alpha(x,t) = 4(n-1)^2\alpha(\alpha-1)P_h^\alpha(x,t). \tag{5.5.2}$$

Thus $P_h^\alpha(x,t)$ is an eigenfunction of Δ_h with eigenvalue $4(n-1)^2\alpha(\alpha-1)$.

Definition 5.5.1 *For $\lambda \in \mathbb{R}$, let $\mathcal{H}_\lambda$ denote the space of all $f \in C^2(\mathbb{B})$ that satisfy*

$$\Delta_h f = \lambda f.$$

As a consequence of the above we have the following analogue of [72, Theorem 4.2.2].

Theorem 5.5.2 *If α and λ are related by $\lambda = 4(n-1)^2\alpha(\alpha-1)$, then $\mathcal{H}_\lambda$ contains every f of the form*

$$f(x) = \int_{\mathbb{S}} P_h^\alpha(x,t)\,d\mu(t), \quad x \in \mathbb{B}, \tag{5.5.3}$$

where μ is a signed Borel measure on $\mathbb{S}$. In particular, $\mathcal{H}_\lambda$ contains the radial function g_α defined by

$$g_\alpha(x) = \int_{\mathbb{S}} P_h^\alpha(x,t)d\sigma(t), \quad x \in \mathbb{B}. \tag{5.5.4}$$

We now proceed to prove the following analogues of [72, Theorems 4.2.3, 4.2.4]. The proofs, with minor changes, follow those given by W. Rudin in [72] for eigenfunction of $\widetilde{\Delta}$. In the following two theorems we will assume that α and λ are related by

$$\lambda = 4(n-1)^2\alpha(\alpha-1).$$

Theorem 5.5.3 *If $f \in \mathcal{H}_\lambda$ and f is radial, then $f(x) = f(0)g_\alpha(x)$.*

Proof. Let $f \in C^2(\mathbb{B})$ be a radial function satisfying $\Delta_h f = \lambda f$. Write $f(x) = u(|x|^2)$. Then by (4.1.3)

$$\Delta_h f(x) = 4r^2(1-r^2)^2u''(r^2) + 2n(1-r^2)^2u'(r^2) + 4(n-2)r^2(1-r^2)u'(r^2).$$

Hence the equation $\Delta_h f = \lambda f$ converts to

$$Lu = \lambda u, \tag{5.5.5}$$

where

$$L(u)(t) = a(t)u''(t) + b(t)u'(t),$$

with

$$a(t) = 4t(1-t)^2 \quad \text{and} \quad b(t) = 2n(1-t)^2 + 4(n-2)t(1-t). \tag{5.5.6}$$

The equation $L(u) = 0$ has a solution (see Exercise 5.7.5(a))

$$u_0(t) = \int_t^1 s^{-\frac{1}{2}n}(1-s)^{n-2}, \quad 0 < t < 1, \tag{5.5.7}$$

which is unbounded as $t \to 0$.

Suppose u_λ is a solution of (5.5.5) with $u_\lambda(0) = 1$, and v is a solution of $u_\lambda^2 v' = u_0'$ for small $t > 0$ where $u_\lambda(t) \neq 0$. Set $w = u_\lambda v$. Then $Lw = \lambda w$ (see Exercise 5.7.5(b)). Since v is unbounded as $t \to 0$, we also have that w is unbounded at 0. Since the solution space of (5.5.5) is two dimensional, the solutions of (5.5.5) that are bounded as $t \to 0$ form a space of dimension one. Since g_α is a bounded solution of $\Delta_h f = \lambda f$, every f in $C^2(\mathbb{B}) \cap \mathcal{H}_\lambda$ satisfies $f(x) = a g_\alpha(x)$ for some constant a. Since $g_\alpha(0) = 1$ the result follows. □

Corollary 5.5.4 $g_\alpha = g_{1-\alpha}$.

Corollary 5.5.4 follows since the definition of λ is unchanged if α is replaced by $1-\alpha$.

Theorem 5.5.5 *Every $f \in \mathcal{H}_\lambda$ satisfies*

$$\int_{\mathbb{S}} f(\psi(rt))d\sigma(t) = g_\alpha(r\eta)f(\psi(0)) \tag{5.5.8}$$

for every $\psi \in \mathcal{M}(\mathbb{B})$, $0 \le r < 1$, $\eta \in \mathbb{S}$. Conversely, if $f \in C(\mathbb{B})$ and f satisfies (5.5.8), then $f \in C^\infty(\mathbb{B})$ and $f \in \mathcal{H}_\lambda$.

Proof. Suppose $f \in \mathcal{H}_\lambda$ and let $f^\sharp$ be the radialization of f given by (4.1.1). Since $f^\sharp \in \mathcal{H}_\lambda$ we have $f^\sharp = f(0)g_\alpha$. If we replace f by $f \circ \psi$, $\psi \in \mathcal{M}(\mathbb{B})$, we obtain (5.5.8).

Conversely, suppose $f \in C(\mathbb{B})$ satisfies (5.5.8). Let $h \in C_c^\infty(\mathbb{B})$ be a radial function such that

$$\int_{\mathbb{B}} h(y) g_\alpha(y) d\tau(y) = 1.$$

In (5.5.8) take $\psi = \varphi_x$. Then with $r = |y|$, by (3.2.5)

$$\int_{O(n)} f(\varphi_x(Ay)) dA = \int_{\mathbb{S}} f(\varphi_x(r\zeta)) d\sigma(\zeta) = g_\alpha(y) f(x). \tag{5.5.9}$$

Multiplying the above by $h(y)d\tau(y)$ and integrating gives

$$f(x) = \int_{\mathbb{B}} f(x) g_\alpha(y) h(y) d\tau(y) = \int_{\mathbb{B}} \int_{O(n)} f(\varphi_x(Ay)) h(y) dA d\tau(y),$$

which by Fubini's theorem and the change of variable $w = \varphi_x(Ay)$

$$= \int_{O(n)} \int_{\mathbb{B}} h(A^{-1}\varphi_x(w)) f(w) d\tau(w) dA.$$

Thus since h is radial, we have

$$f(x) = \int_{\mathbb{B}} f(w) h(\varphi_x(w)) d\tau(w).$$

As a consequence, f is C^∞ on $\mathbb{B}$.

If in (5.5.9) we compute Δ_h with respect to y we obtain

$$\begin{aligned} \lambda g_\alpha(y) f(x) &= (\Delta_h)_y [g_\alpha(y) f(x)] \\ &= \int_{O(n)} (\Delta_h)_y [f \circ \varphi_x \circ A](y) dA. \end{aligned}$$

But by the invariance of Δ_h,

$$(\Delta_h)_y [f \circ \varphi_x \circ A](y) = (\Delta_h f)(\varphi_x(Ay)).$$

Therefore

$$\lambda g_\alpha(y) f(x) = \int_{O(n)} (\Delta_h f)(\varphi_x(Ay)) dA.$$

Setting $y = 0$ gives

$$\lambda f(x) = \int_{O(n)} (\Delta_h f)(\varphi_x(0)) dA = \Delta_h f(x),$$

which proves the result. □

The following corollary provides another proof of Corollary 4.5.5.

Corollary 5.5.6 *If f is a continuous function on $\mathbb{B}$ satisfying*

$$f(\psi(0)) = \int_{\mathbb{S}} f(\psi(rt))d\sigma(t)$$

for all $\psi \in (\mathbb{B})$ and $0 < r < 1$, then $f \in C^\infty(\mathbb{B})$ with $\Delta_h f = 0$.

We close this section by obtaining asymptotic estimates for the function g_α. By Equation (5.5.4) the function g_α is given by

$$g_\alpha(x) = (1 - |x|^2)^{\alpha(n-1)} \int_{\mathbb{S}} \frac{d\sigma(\zeta)}{|x - \zeta|^{2\alpha(n-1)}}. \tag{5.5.10}$$

For $\gamma > 0$, $0 < r < 1$, and $t \in \mathbb{S}$, set

$$I_\gamma(r) = \int_{\mathbb{S}} |rt - \zeta|^{-\gamma} d\sigma(\zeta).$$

The following asymptotic estimates for I_γ appear to be well known, but are usually stated without a reference to a proof.

Theorem 5.5.7 *For $0 < r < 1$ and $\gamma > 0$,*

$$I_\gamma(r) \approx \begin{cases} 1 & \text{if } \gamma < (n-1), \\ \log \frac{1}{(1-r)} & \text{if } \gamma = (n-1), \\ (1-r)^{-\gamma+(n-1)} & \text{if } \gamma > (n-1). \end{cases}$$

Proof. Since the result is well known when $n = 2$ ([72, Proposition 1.4.10]), we will assume that $n \geq 3$. Set $e_1 = (1, 0')$ where $0'$ is the zero in $\mathbb{R}^{n-1}$. Then since σ is invariant under $O(n)$,

$$I_\gamma(r) = \int_{\mathbb{S}} |re_1 - \zeta|^{-\gamma} d\sigma(\zeta),$$

which by writing $\zeta = (\zeta_1, \zeta')$, $\zeta' \in \mathbb{R}^{n-1}$,

$$= \int_{\mathbb{S}} \frac{d\sigma(\zeta)}{[|r - \zeta_1|^2 + |\zeta'|^2]^{\gamma/2}}.$$

Consider the map $\Psi : (-1, 1) \times \mathbb{S}_{n-1} \to \mathbb{S}_n$ defined by

$$\Psi(x, \zeta) = (x, \sqrt{1 - x^2}\,\zeta).$$

The map Ψ is one-to-one and range $\Psi = \mathbb{S}_n \setminus \{(x, \zeta) : \zeta = 0\}$. By [10, Appendix A.4],

$$d\sigma_n(\Psi(x, \zeta)) = (1 - x^2)^{\frac{n-3}{2}} dx\, d\sigma_{n-1}(\zeta)$$

and

$$\int_{\mathbb{S}_n} f d\sigma_n = c_n \int_{-1}^{1} (1-x^2)^{\frac{n-3}{2}} \int_{\mathbb{S}_{n-1}} f(x, \sqrt{1-x^2}\,\zeta) d\sigma_{n-1}(\zeta) dx, \tag{5.5.11}$$

where c_n is a constant depending only on n. Therefore

$$I_\gamma(r) = c_n \int_{-1}^{1} \frac{(1-x^2)^{\frac{n-3}{2}} dx}{[|r-x|^2 + (1-x^2)]^{\gamma/2}}.$$

Since $(r-x)^2 + (1-x^2) = 1 - 2rx + r^2$

$$I_\gamma(r) = c_n \int_{-1}^{1} \frac{(1-x^2)^{\frac{n-3}{2}} dx}{[1 - 2rx + r^2]^{\gamma/2}}. \tag{5.5.12}$$

With the change of variable $x = 1 - t$,

$$I_\gamma(r) = c_n \int_0^2 \frac{t^{\frac{n-3}{2}} (2-t)^{\frac{n-3}{2}}}{[(1-r)^2 + 2rt]^{\gamma/2}} dt.$$

Since we are only interested in the behavior of $I_\gamma(r)$ as $r \to 1$, we assume that $r \geq \frac{1}{2}$. Thus since $n \geq 3$ we have

$$\begin{aligned} I_\gamma(r) &\leq 2^{\frac{n-3}{2}} c_n \int_0^2 \frac{t^{\frac{n-3}{2}} dt}{[(1-r)^2 + t]^{\gamma/2}} \\ &\leq C_n \int_0^2 t^{\frac{n-3}{2} - \frac{\gamma}{2}} dt, \end{aligned}$$

which is finite provided $\gamma < (n-1)$.

Suppose now that $\gamma \geq (n-1)$. We now make the change of variable $t = (1-r)^2 w$ to obtain

$$I_\gamma(r) \leq C_n (1-r)^{n-1-\gamma} \int_0^{2/(1-r)^2} \frac{w^{\frac{n-3}{2}} dw}{[1+w]^{\gamma/2}}.$$

Therefore

$$\begin{aligned} I_\gamma(r) &\leq C_n (1-r)^{n-1-\gamma} \left\{ \int_0^1 (1+w)^{-\gamma/2} dw + \int_1^{2/(1-r)^2} w^{\frac{1}{2}(n-3-\gamma)} dw \right\} \\ &\leq C_n (1-r)^{n-1-\gamma} \left\{ C_\gamma + \int_1^{2/(1-r)^2} w^{\frac{1}{2}(n-3-\gamma)} dw \right\}. \end{aligned}$$

If $\gamma = (n-1)$, then

$$I_\gamma(r) \leq C_n \left\{ C_\gamma + \int_1^{2/(1-r)^2} \frac{1}{w} dw \right\} \leq C_{n,\gamma} \log \frac{1}{(1-r)}.$$

If $\gamma > (n-1)$, then $\frac{1}{2}(n-3-\gamma) < -1$ and thus

$$I_\gamma(r) \le C_n(1-r)^{n-1-\gamma}\left\{C_\gamma + \int_1^\infty w^{\frac{1}{2}(n-3-\gamma)}dw\right\} \le C_{n,\gamma}(1-r)^{n-1-\gamma}.$$

For the lower estimate we have that

$$\begin{aligned} I_\gamma(r) &= c_n \int_0^2 \frac{t^{\frac{n-3}{2}}(2-t)^{\frac{n-3}{2}}}{[(1-r)^2+2rt]^{\gamma/2}}dt \\ &\ge \frac{c_n}{2^{\gamma/2}}\int_0^1 t^{\frac{1}{2}(n-3-\gamma)}dt. \end{aligned}$$

If $\gamma < (n-1)$, then the above integral is finite. For $\gamma \ge (n-1)$ we have

$$I_\gamma(r) \ge C_{n,\gamma}\int_{(1-r)}^1 t^{\frac{1}{2}(n-3-\gamma)}dt,$$

from which the result follows. □

Combining the above with (5.5.10) gives the following corollary.

Corollary 5.5.8

$$g_\alpha(x) \approx \begin{cases} (1-|x|^2)^{\alpha(n-1)} & \textit{if } \alpha < \frac{1}{2}, \\ (1-|x|^2)^{\frac{1}{2}(n-1)}\log\frac{1}{(1-|x|^2)} & \textit{if } \alpha = \frac{1}{2}, \\ (1-|x|^2)^{(1-\alpha)(n-1)} & \textit{if } \alpha > \frac{1}{2}. \end{cases}$$

As a consequence of the above we have

(a) If $0 < \alpha \le \frac{1}{2}$, then g_α is continuous on $\mathbb{B}$ with $\lim_{|x|\to 1} g_\alpha(x) = 0$, that is, $g_\alpha \in C_0(\mathbb{B})$.

(b) If $\alpha = 0$, then g_α is bounded on $\mathbb{B}$ but $g_\alpha \notin C_0(\mathbb{B})$.

(c) If $\alpha < 0$, then g_α is unbounded.

5.6 The Poisson Kernel on $\mathbb{H}$

Recall that the upper half-space $\mathbb{H}$ in $\mathbb{R}^n$ is the set

$$\mathbb{H} = \{z = (x, y) \in \mathbb{R}^n : x \in \mathbb{R}^{n-1}, y > 0\}.$$

As is usual we identify $\mathbb{R}^{n-1}$ with $\mathbb{R}^{n-1} \times \{0\}$. With this convention we have that $\partial\mathbb{H} = \mathbb{R}^{n-1}$.

For $z = (x, y) \in \mathbb{H}$, $x \in \mathbb{R}^{n-1}$, $y > 0$, and $t \in \mathbb{R}^{n-1}$, set

$$P_{\mathbb{H}}(z, t) = c_n\left[\frac{y}{|x-t|^2 + y^2}\right]^{n-1}, \tag{5.6.1}$$

where c_n is chosen so that $\int_{\mathbb{R}^{n-1}} P_{\mathbb{H}}(z,t)dt = 1$. The function $P_{\mathbb{H}}$ is called the Poisson kernel for the upper half-space $\mathbb{H}$. As a consequence of Exercises 5.7.16 and 5.7.17 it follows that for fixed $t \in \mathbb{R}^{n-1}$ the function $x \to P_{\mathbb{H}}(x,t)$ is $\mathcal{H}$-harmonic on $\mathbb{H}$.

For $y > 0$ set

$$v_y(x) = c_n \left[\frac{y}{|x|^2 + y^2}\right]^{n-1}.$$

The function v_y has the following property: $v_1(\frac{x}{y}) = y^{n-1}v_y(x)$. Consider $\int_{\mathbb{R}^{n-1}} v_y(x)dx$, where dx denotes Lebesgue measure on $\mathbb{R}^{n-1}$. By the change of variable $x = yx'$ we have

$$\begin{aligned}\int_{\mathbb{R}^{n-1}} v_y(x)dx &= \int_{\mathbb{R}^{n-1}} v_y(yx')y^{n-1}dx' \\ &= \int_{\mathbb{R}^{n-1}} v_1(x')dx' = c_n \int_{\mathbb{R}^{n-1}} \frac{dx}{(|x|^2+1)^{n-1}}.\end{aligned}$$

Hence $\int v_y(x)dx$ is independent of y. Since $\int v_1(x)dx < \infty$, we can choose c_n such that

$$\int_{\mathbb{R}^{n-1}} v_y(x)dx = c_n \int_{\mathbb{R}^{n-1}} \frac{dx}{(|x|^2+1)^{n-1}} = 1. \tag{5.6.2}$$

By expressing the integral in (5.6.2) in polar coordinates it is easily shown that

$$c_n = \frac{2\Gamma(n-1)}{(n-1)V(\mathbb{B}_{n-1})\Gamma^2(\frac{1}{2}(n-1))},$$

where Γ is the gamma function (Exercise 5.7.11).

The above computations show that $P_{\mathbb{H}}(z,t)$ is positive, $\mathcal{H}$-harmonic on $\mathbb{H}$, and with the above choice of c_n satisfies

$$\int_{\mathbb{R}^{n-1}} P_{\mathbb{H}}(z,t)\,dt = 1, \quad z = (x,y) \in \mathbb{H}. \tag{5.6.3}$$

The next result proves that $P_{\mathbb{H}}$ is an approximate identity on $\mathbb{H}$.

Lemma 5.6.1 *For every $a \in \mathbb{R}^{n-1}$ and $\delta > 0$,*

$$\lim_{z\to a} \int_{|t-a|>\delta} P_{\mathbb{H}}(z,t)dt = 0.$$

Proof. We prove the result for $n \geq 3$, with the obvious modifications when $n = 2$. Let $\delta > 0$ be given. Since $P_{\mathbb{H}}(z,t) \leq c_n y^{n-1}|x-t|^{-2(n-1)}$, for $|x-a| < \frac{\delta}{2}$, we have $P_{\mathbb{H}}(x,t) \leq C_n y^{n-1}|a-t|^{-2(n-1)}$. Therefore

$$\int_{|a-t|>\delta} P_{\mathbb{H}}(z,t)\,dt \leq C_n y^{n-1} \int_{|a-t|>\delta} \frac{dt}{|a-t|^{2(n-1)}}.$$

Expressing the above integral in polar coordinates yields

$$\int_{|a-t|>\delta} \frac{dt}{|a-t|^{2(n-1)}} = \frac{C_n}{(n-2)\delta^{n-2}},$$

from which the result now follows. □

For $1 \le p \le \infty$ let $L^p(\mathbb{R}^{n-1})$ denote the space of Lebesgue measurable functions on $\mathbb{R}^{n-1}$ for which $\|f\|_p < \infty$, where for $1 \le p < \infty$

$$\|f\|_p = \left[\int_{\mathbb{R}^{n-1}} |f(x)|^p dx\right]^{1/p},$$

and for $p = \infty$, $\|f\|_\infty$ denotes the essential supremum of f. The Poisson integral of $f \in L^p(\mathbb{R}^{n-1})$, $1 \le p \le \infty$, is the function

$$P_{\mathbb{H}}[f](z) = \int_{\mathbb{R}^{n-1}} P_{\mathbb{H}}(z,t)f(t)dt.$$

Since $P_{\mathbb{H}}(z,\cdot) \in L^q(\mathbb{R}^{n-1})$ for every $q \in [1,\infty]$ (Exercise 5.7.13), $P_{\mathbb{H}}[f](z)$ is well defined for every $z \in \mathbb{H}$.

Theorem 5.6.2 (Dirichlet problem for $\mathbb{H}$) *Suppose f is continuous and bounded on $\mathbb{R}^{n-1}$. For $z = (x,y) \in \overline{\mathbb{H}}$, define F on $\overline{\mathbb{H}}$ by*

$$F(z) = \begin{cases} P_{\mathbb{H}}[f](z), & z \in \mathbb{H}, \\ f(x), & x \in \mathbb{R}^{n-1}. \end{cases}$$

Then F is continuous on $\overline{\mathbb{H}}$ and $\mathcal{H}$-harmonic on $\mathbb{H}$ with $|F(z)| \le \|f\|_\infty$ for all $z \in \mathbb{H}$.

Proof. That $|F(z)| \le \|f\|_\infty$ is an immediate consequence of (5.6.3). Let $a \in \mathbb{R}^{n-1}$ and $\delta > 0$. Then

$$\begin{aligned} |F(z) - f(a)| &= \left|\int_{\mathbb{R}^{n-1}} P_{\mathbb{H}}(z,t)(f(t) - f(a))dt\right| \\ &\le \int_{|t-a|\le\delta} P_{\mathbb{H}}(z,t)|f(t) - f(a)|dt + 2\|f\|_\infty \int_{|t-a|>\delta} P_{\mathbb{H}}(z,t)dt \end{aligned}$$

for all $z \in \mathbb{H}$. The result now follows as in Theorem 5.3.3 from the continuity of f at a and Lemma 5.6.1. □

5.7 Exercises

5.7.1. (a) If $\hat{f} \in L^p$, $1 \le p < \infty$, prove that

$$\lim_{r\to 1} \int_{\mathbb{S}} |P_h[\hat{f}](rt) - \hat{f}(t)|^p d\sigma(t) = 0.$$

(b) If $\hat{f} \in L^\infty(\mathbb{S})$, prove that

$$\lim_{r\to 1} \int_{\mathbb{S}} P_h[\hat{f}](rt)g(t)d\sigma(t) = \int_{\mathbb{S}} \hat{f}(t)g(t)d\sigma(t)$$

for all $g \in L^1(\mathbb{S})$.

5.7.2. Suppose U is a positive $\mathcal{H}$-harmonic function on $\mathbb{B}$ and that U extends continuously to $\overline{\mathbb{B}} \setminus \{\zeta\}$, $\zeta \in \mathbb{S}$, with $U = 0$ on $\mathbb{S} \setminus \{\zeta\}$. Prove that $U(x) = cP_h(x,\zeta)$ for some $c > 0$.

5.7.3. Prove that

$$|\nabla_x P_h(x,t)|^2 = 4(n-1)^2 \frac{(1-|x|^2)^{2(n-2)}}{|x-t|^{4(n-1)}}.$$

5.7.4. If $f \in \mathcal{H}_\lambda$, $\lambda \geq 0$, prove that $|f|^p$ is $\mathcal{H}$-subharmonic for all p, $1 \leq p < \infty$.

5.7.5. (a) Prove that $u_0(t)$ given by (5.5.7) is a solution of

$$au'' + bu' = 0$$

where a and b are given by (5.5.6).

(b) Let u_λ and v be as given in the proof of Theorem 5.5.3, and let $w = u_\lambda v$. Prove that $Lw = \lambda w$.

5.7.6. Let $\{f_n\}$ be a sequence in $\mathcal{H}_\lambda$ that converges to f uniformly on compact subsets of $\mathbb{B}$. Prove that $f \in \mathcal{H}_\lambda$.

5.7.7. Given α, $-\infty < \alpha \leq \frac{1}{2}$, determine all values of p such that $g_\alpha \in L^p(\mathbb{B},\tau)$.

5.7.8. Suppose $f(x) = P_h[\hat{f}](x)$ where $\hat{f} \in L^1(\mathbb{S})$. Prove that

$$|\nabla^h f(x)| \leq 2(n-1)P_h[|\hat{f}|](x),$$

5.7.9. For $f \in L^1(\mathbb{S})$, let $u(x) = P_h[f](x)$ and $v(x) = P_e[f](x)$. Show that

$$v(x) = \int_0^1 \eta(|x|,s)u(sx)ds,$$

where η is given by (5.2.1).

5.7.10. **One radius theorem** [72, Theorem 4.3.4]. The following theorem is true for $\mathcal{M}$-harmonic functions on the unit ball in $\mathbb{C}^n$.

Theorem *Suppose $u \in C(\overline{\mathbb{B}})$ and suppose that to every $z \in \mathbb{B}$ there corresponds one radius $r(z)$, $0 < r(z) < 1$, such that*

$$u(z) = \int_{\mathbb{S}} u(\varphi_z(r(z)\zeta)d\sigma(\zeta).$$

Then u is $\mathcal{M}$-harmonic in $\mathbb{B}$.

The analogue of this result for Euclidean harmonic functions may be found in [10].

Question (*): Is the analogous result true for $\mathcal{H}$-harmonic functions in $\mathbb{B}$?

Exercises on the Upper Half-Space $\mathbb{H}$

5.7.11. Compute the value of c_n in Equation (5.6.2).

5.7.12. Prove Lemma 5.6.1.

5.7.13. Prove that $P_{\mathbb{H}}(z,\cdot) \in L^q(\mathbb{R}^{n-1})$ for every $q \in [1,\infty]$.

For a function U on $\mathbb{H}$ and $y > 0$, we let U_y denote the function on $\mathbb{R}^{n-1}$ defined by

$$U_y(x) = U(x,y).$$

The functions U_y play the same role on $\mathbb{H}$ that the dilations f_r, $0 < r < 1$, play on the unit ball $\mathbb{B}$.

5.7.14. Suppose $1 \le p < \infty$. If $f \in L^p(\mathbb{R}^{n-1})$ and $U = P_{\mathbb{H}}[f]$, prove that

(a) $\|U_y\|_p \le \|f\|_p$ for all $y > 0$,

(b) $|U_y(x)| \le C\dfrac{\|f\|_p}{y^{(n-1)/p}}$, and

(c) $\lim_{y\to 0} \|U_y - f\|_p = 0$.

5.7.15. For $z = (x,y) \in \mathbb{H}$, $x \in \mathbb{R}^{n-1}$, $y > 0$, let $\Phi : \mathbb{H} \to \mathbb{B}$ be the mapping given by (2.3.2), that is,

$$\Phi(z) = -e_n + \frac{2(z+e_n)}{|z+e_n|^2},$$

where $e_n = (0,1)$, $0 \in \mathbb{R}^{n-1}$. For $w \in \mathbb{B}$, consider

$$U(w) = \frac{1-|w|^2}{|e_n - w|^2}.$$

Show that

$$U(\Phi(z)) = \frac{y}{|x|^2+y^2}.$$

5.7.16. For $x \in \mathbb{R}^{n-1}$, $y > 0$, set

$$V(x,y) = \frac{y}{|x|^2+y^2}.$$

Establish each of the following:

(a) $|\nabla V|^2 = \dfrac{1}{(|x|^2+y^2)^2}$.

(b) $\Delta V = \dfrac{-2y(n-2)}{(|x|^2+y^2)^2}$.

5.7.17. Let $h(x,y) = (V(x,y))^{n-1}$ and let

$$L_{\mathbb{H}} = y^2\Delta h - (n-2)y\frac{\partial h}{\partial y}$$

be the invariant Laplacian on $\mathbb{H}$.

(a) Prove that

$$L_{\mathbb{H}} = y(n-1)V^{n-3}\left[y(n-2)|\nabla V|^2 + yV\Delta V - (n-2)V\frac{\partial V}{\partial y}\right].$$

(b) Using the results of the previous exercise show that $Lh = 0$ on $\mathbb{H}$.

5.7.18. Fix $a \in \mathbb{R}^{n-1}$. Let U be a positive $\mathcal{H}$-harmonic function on $\mathbb{H}$ that extends continuously to $\overline{\mathbb{H}} \setminus \{a\}$ with boundary values 0 on $\mathbb{R}^{n-1} \setminus \{a\}$. Suppose also that

$$\lim_{y\to\infty} \frac{U(0,y)}{y^{n-1}} = 0.$$

Prove that $U(x) = cP_{\mathbb{H}}(x, a)$ for some $c > 0$.

5.7.19. Suppose that U is a positive $\mathcal{H}$-harmonic function on $\mathbb{H}$, continuous on $\overline{\mathbb{H}}$ with boundary value 0 on $\mathbb{R}^{n-1}$. Prove that there exists a constant $c > 0$ such that $U(x, y) = cy^{n-1}$ for all $(x, y) \in \mathbb{H}$.

6

Spherical Harmonic Expansions

In this chapter we provide a brief survey of spherical harmonic and zonal harmonic functions in order to obtain the zonal harmonic expansion of the Poisson kernel P_h and the spherical harmonic expansion of $\mathcal{H}$-harmonic functions. For details concerning the general theory of spherical harmonics the reader is referred to the text [10] by S. Axler, P. Bourdon, and W. Ramey.

Throughout this chapter we will assume $n \geq 3$. The results for $n = 2$ are well known. As in [10, Chapter 5], for $m = 0, 1, 2, \ldots$, we denote by $\mathcal{H}_m(\mathbb{R}^n)$ the space of all (Euclidean) **homogeneous**[1] **harmonic polynomials of degree** m on $\mathbb{R}^n$. A **spherical harmonic of degree** m is the restriction to $\mathbb{S}$ of a harmonic polynomial in $\mathcal{H}_m(\mathbb{R}^n)$. The collection of all spherical harmonic polynomials of degree m will be denoted by $\mathcal{H}_m(\mathbb{S})$. Every element of $\mathcal{H}_m(\mathbb{S})$ has a unique extension to $\mathcal{H}_m(\mathbb{R}^n)$. If $m \neq k$, then $\mathcal{H}_m(\mathbb{S})$ and $\mathcal{H}_k(\mathbb{S})$ are orthogonal in $L^2(\mathbb{S})$, that is,

$$\langle p, q \rangle = \int_{\mathbb{S}} p(t)q(t)\, d\sigma(t) = 0$$

for all $p \in \mathcal{H}_m(\mathbb{S})$, $q \in \mathcal{H}_k(\mathbb{S})$. Furthermore,

$$L^2(\mathbb{S}) = \bigoplus_{m=0}^{\infty} \mathcal{H}_m(\mathbb{S}),$$

that is, for each $f \in L^2(\mathbb{S})$, there exists $p_m \in \mathcal{H}_m(\mathbb{S})$ such that

$$f = \sum_{m=0}^{\infty} p_m, \tag{6.0.1}$$

where the series converges in $L^2(\mathbb{S})$.

[1] A polynomial p on $\mathbb{R}^n$ is **homogeneous** of degree m if $p(\lambda x) = \lambda^m p(x)$ for all $\lambda \in \mathbb{R}$.

6.1 Dirichlet Problem for Spherical Harmonics

Our first goal is to solve the Dirichlet problem for $p_\alpha \in \mathcal{H}_\alpha(\mathbb{S})$. Since each p_α has a unique extension to $\mathcal{H}_\alpha(\mathbb{R}^n)$, we can assume that $p_\alpha(x) \in \mathcal{H}_\alpha(\mathbb{R}^n)$. Set $f(x) = g(r^2)p_\alpha(x)$, where $r^2 = |x|^2$. Since $\langle x, \nabla p_\alpha\rangle = \alpha p_\alpha(x)$, we have

$$\begin{aligned}\langle x, \nabla f(x)\rangle &= p_\alpha(x)\langle x, \nabla g\rangle + g(r^2)\langle x, \nabla p_\alpha\rangle \\ &= 2r^2 g'(r^2)p_\alpha(x) + \alpha g(r^2)p_\alpha(x),\end{aligned}$$

and since p_α is harmonic,

$$\begin{aligned}\Delta f(x) &= 2\langle \nabla p_\alpha, \nabla g\rangle + p_\alpha(x)\Delta g(r^2) \\ &= 4\alpha p_\alpha(x)g'(r^2) + p_\alpha(x)[2ng'(r^2) + 4r^2 g'(r^2)].\end{aligned}$$

Therefore

$$\begin{aligned}\Delta_h f(x) &= 2(1-r^2)p_\alpha(x) \\ &\times \left[2(1-r^2)r^2 g''(r^2) + \{(n+2\alpha)(1-r^2) + 2r^2(n-2)\}g'(r^2) + \alpha(n-2)g(r^2)\right].\end{aligned}$$

Thus in order that $\Delta_h f(x) = 0$ we must have

$$2(1-r^2)r^2 g''(r^2) + \{(n+2\alpha)(1-r^2) + 2r^2(n-2)\}g'(r^2) + \alpha(n-2)g(r^2) = 0,$$

or

$$(1-r^2)r^2 g''(r^2) + \{(\alpha + \tfrac{1}{2}n) - (\alpha + 2 - \tfrac{1}{2}n)r^2\}g'(r^2) - \alpha(1 - \tfrac{1}{2}n)g(r^2) = 0.$$

If we set $t = r^2$, $a = \alpha$, $b = 1 - \frac{1}{2}n$, and $c = \alpha + \frac{1}{2}n$, then the above equation can be rewritten as

$$t(1-t)g''(t) + \{c - (a+b+1)t\}g'(t) - abg(t) = 0. \tag{6.1.1}$$

Equation (6.1.1), however, is the **hypergeometric equation**, for which a particular solution is given by the **hypergeometric function** $F(a,b;c;t)$ [1, Identity 15.5.1], [20, Chapter II], [50] defined by

$$F(a,b;c;z) = \sum_{k=0}^{\infty} \frac{(a)_k(b)_k}{(c)_k}\frac{z^k}{k!}, \qquad |z| < 1 \tag{6.1.2}$$

In the above, $(a)_0 = 1$ and for $k = 1, 2, \ldots,$

$$(a)_k = a(a+1)\cdots(a+k-1).$$

If a is not a negative integer, then

$$(a)_k = \Gamma(a+k)/\Gamma(a),$$

where Γ is the gamma function defined on $\mathbb{C}\setminus\{0,-1,-2,...\}$. If $c-a-b>0$, then the series in (6.1.2) converges absolutely for all z, $|z|\le 1$. For the above values of a, b, and c, we have $c-a-b=n-1$. Thus the function $g(r^2)$ is given by

$$g(r^2)=c_\alpha F(\alpha, 1-\tfrac{1}{2}n; \alpha+\tfrac{1}{2}n; r^2), \tag{6.1.3}$$

for an arbitrary constant c_α. Define $S_{n,\alpha}(r)$ by

$$S_{n,\alpha}(r)=\frac{F(\alpha,1-\frac{1}{2}n;\alpha+\frac{1}{2}n;r^2)}{F(\alpha,1-\frac{1}{2}n;\alpha+\frac{1}{2}n;1)}. \tag{6.1.4}$$

Then $S_{n,\alpha}(1)=1$ and $f(x)=S_\alpha(|x|)p_\alpha(x)$ is a solution of $\Delta_h f(x)=0$ that is continuous on $\overline{\mathbb{B}}$ with $f(\zeta)=p_\alpha(\zeta)$ for all $\zeta\in\mathbb{S}$. This proves the following theorem.

Theorem 6.1.1 *If $p_\alpha\in\mathcal{H}_\alpha(\mathbb{S})$, $\alpha=0,1,2,\dots$, then for all $t\in\mathbb{S}$,*

$$P_h[p_\alpha](rt)=r^\alpha S_{n,\alpha}(r)p_\alpha(t),$$

where $S_{n,\alpha}$ is defined by (6.1.4).

Example 6.1.2 If n is even, say $n=2m$, then $b=1-m$ and thus $(b)_k=0$ for all $k\ge m$. Hence $g(r^2)$ is a polynomial of degree $n-2$. When $n=4$, $b=-1$ and

$$S_{4,\alpha}(r)=\tfrac{1}{2}(\alpha+2)\left(1-\frac{\alpha}{\alpha+2}r^2\right), \qquad \alpha=0,1,2,\dots.$$

When $n=6$, $b=-2$ and

$$F(\alpha,-2;\alpha+3;r^2)=1-\frac{2\alpha}{\alpha+3}r^2+\frac{\alpha(\alpha+1)}{(\alpha+3)(\alpha+4)}r^4.$$

Thus for $\alpha=0,1,2,\dots,$

$$S_{6,\alpha}(r)=\tfrac{1}{12}(\alpha+3)(\alpha+4)\left[1-\frac{2\alpha}{\alpha+3}r^2+\frac{\alpha(\alpha+1)}{(\alpha+3)(\alpha+4)}r^4\right].$$

By [50, Identity 9.3.4], if $c-a-b>0$ then

$$\lim_{t\to 1^-}F(a,b;c;t)=\frac{\Gamma(c)\Gamma(c-a-b)}{\Gamma(c-a)\Gamma(c-b)}. \tag{6.1.5}$$

Therefore

$$F(\alpha,1-\tfrac{1}{2}n;\alpha+\tfrac{1}{2}n;1)=\frac{\Gamma(\alpha+\frac{1}{2}n)\Gamma(n-1)}{\Gamma(\frac{1}{2}n)\Gamma(\alpha+n-1)},$$

and hence $S_{n,\alpha}(r) = c_{n,\alpha} F(\alpha, 1 - \frac{1}{2}n; \alpha + \frac{1}{2}n; r^2)$, where

$$c_{n,\alpha} = \frac{\Gamma(\frac{1}{2}n)\Gamma(\alpha + n - 1)}{\Gamma(\alpha + \frac{1}{2}n)\Gamma(n - 1)}. \tag{6.1.6}$$

Also, using the transformation [1, Identity 15.3.3], [50, Identity 9.5.3],

$$F(a, b; c; t) = (1 - t)^{c-a-b} F(c - a, c - b; c; t),$$

we can express $S_{n,\alpha}(r)$ as

$$S_{n,\alpha}(r) = c_{n,\alpha}(1 - r^2)^{n-1} F(\tfrac{1}{2}n, \alpha + n - 1; \alpha + \tfrac{1}{2}n; r^2). \tag{6.1.7}$$

Theorem 6.1.1 can be used to compute the invariant Poisson integral of a polynomial q on $\mathbb{S}$. By [10, Corollary 5.7], if q is a polynomial on $\mathbb{R}^n$ of degree m, then the restriction of q to $\mathbb{S}$ is a sum of spherical harmonics of degree at most m. That is, there exist $p_k \in \mathcal{H}_k(\mathbb{S})$, $k = 0, 1, \ldots, m$, such that $q(t) = \sum_{k=0}^{m} p_k(t)$ for all $t \in \mathbb{S}$. Hence

$$P_h[q](x) = \sum_{k=0}^{m} P_h[p_k](x).$$

But by Theorem 6.1.1, $P_h[p_k](x) = S_{n,k}(|x|)p_k(x)$. Thus

$$P_h[q](x) = \sum_{k=0}^{m} S_{n,k}(|x|)p_k(x).$$

The above computations are particularly easy when n is even. These computations are illustrated in the following examples.

Examples 6.1.3 **(a)** For our first example we consider the function $q(t) = t_1^2$ in $\mathbb{R}^4$. Then for $t \in \mathbb{S}$, $q(t) = p_0(t) + p_2(t)$, where $p_0(x) = \frac{1}{4}$ and $p_2(x) = x_1^2 - \frac{1}{4}|x|^2$. Thus by Example 6.1.2, in $\mathbb{R}^4$

$$\begin{aligned} P_h[t_1^2](x) &= \tfrac{1}{4} + S_{4,2}(|x|)p_2(x) \\ &= \tfrac{1}{4} + (2 - |x|^2)(x_1^2 - \tfrac{1}{4}|x|^2). \end{aligned}$$

The above function is easily shown to be $\mathcal{H}$-harmonic on $\mathbb{B}$.

(b) In spaces of odd dimension these computations are much more complicated. As in Example 5.3.7 consider $q(t) = t_1^2$ in $\mathbb{R}^3$. Then $q(t) = p_0(t) + p_2(t)$ where $p_0(x) = \frac{1}{3}$ and $p_2(x) = x_1^2 - \frac{1}{3}|x|^2$. Hence

$$P_h[q](x) = \tfrac{1}{3} + S_{3,2}(|x|)(x_1^2 - \tfrac{1}{3}|x|^2).$$

Unfortunately, however, there is no simple expression for $S_{3,2}(r)$. The function $S_{3,2}(r)$ is given by

$$\begin{aligned} S_{3,2}(r) &= c_{3,2}F(2, -\tfrac{1}{2}; \tfrac{7}{2}; r^2) \\ &= \frac{\Gamma(\frac{3}{2})\Gamma(4)}{\Gamma(-\frac{1}{2})} \sum_{k=0}^{\infty} \frac{\Gamma(k+2)\Gamma(k-\frac{1}{2})}{\Gamma(k+\frac{7}{2})k!} r^{2k} \\ &= -\frac{3}{2} \sum_{k=0}^{\infty} \frac{(k+1)\Gamma(k-\frac{1}{2})}{\Gamma(k+\frac{7}{2})} r^{2k}. \end{aligned}$$

Thus

$$P_h[t_1^2](x) = \tfrac{1}{3} - \tfrac{3}{2}(x_1^2 - \tfrac{1}{3}|x|^2) \sum_{k=0}^{\infty} \frac{(k+1)\Gamma(k-\frac{1}{2})}{\Gamma(k+\frac{7}{2})} |x|^{2k}.$$

Using (6.1.7) we also have

$$S_{3,2}(r) = 4(1-r^2)^2 \sum_{k=0}^{\infty} \frac{(k+3)(k+2)(k+1)}{(2k+5)(2k+3)} r^{2k}.$$

6.2 Zonal Harmonic Expansion of the Poisson Kernel

Our next goal is to obtain an expansion of the Poisson kernel P_h in terms of the zonal harmonics. Fix a point $\eta \in \mathbb{S}$. By considering the linear functional $\gamma : \mathcal{H}_m(\mathbb{S}) \to \mathbb{R}$ defined by $\gamma(p) = p(\eta)$, it follows from the Riesz representation theorem that there exists a unique function $Z_\eta^{(m)} \in \mathcal{H}_m(\mathbb{S})$ such that

$$p(\eta) = \int_{\mathbb{S}} p(t) Z_\eta^{(m)}(t)\, d\sigma(t)$$

for all $p \in \mathcal{H}_m(\mathbb{S})$. The spherical harmonic $Z_\eta^{(m)}$ is called the **zonal harmonic** of degree m with pole η. Set $Z_m(\eta, \zeta) = Z_\eta^{(m)}(\zeta)$. It is an easy exercise to show that the zonal harmonic Z_m satisfies

$$Z_m(\eta, \zeta) = Z_m(\zeta, \eta), \tag{6.2.1}$$

$$Z_m(A\eta, A\zeta) = Z_m(\eta, \zeta) \quad \text{for all } A \in O(n), \text{ and} \tag{6.2.2}$$

$$Z_m(\eta, \eta) = \|Z_\eta^{(m)}\|_2^2 = h_m = \dim \mathcal{H}_m(\mathbb{S}). \tag{6.2.3}$$

Also, for fixed $\eta \in \mathbb{S}$, the function $Z_\eta^{(m)}$ has a unique extension to a harmonic function on $\mathbb{R}^n$. This function will again be denoted by $Z_m(x, \eta)$. An explicit formula for Z_m is given in [10, Theorem 5.2.4], which for completeness we state as a theorem without proof.

Theorem 6.2.1 *Let $x \in \mathbb{R}^n$ and let $\eta \in \mathbb{S}$, then*

$$Z_m(x,\eta) = (n+2m-2)\sum_{k=0}^{[m/2]}(-1)^k \frac{n(n+2)\cdots(n+2m-2k-4)}{2^k k!(m-2k)!}\langle x,\eta\rangle^{m-2k}|x|^{2k}.$$

In terms of zonal harmonics, the decomposition (6.0.1) of $L^2(\mathbb{S})$ can now be expressed as follows: for $f \in L^2(\mathbb{S})$

$$f(\eta) = \sum_{\alpha=0}^{\infty}\langle Z_\eta^{(\alpha)}, f\rangle = \sum_{\alpha=0}^{\infty}\int_{\mathbb{S}} Z_\alpha(\eta,\zeta)f(\zeta)d\sigma(\zeta), \tag{6.2.4}$$

where the series converges in $L^2(\mathbb{S})$.

We now derive the following expansion of the invariant Poisson kernel on $\mathbb{B}$ in terms of the zonal harmonics Z_α.

Theorem 6.2.2 *For $x \in \mathbb{B}$, $t \in \mathbb{S}$,*

$$P_h(x,t) = \sum_{\alpha=0}^{\infty}|x|^\alpha S_{n,\alpha}(|x|)Z_\alpha(\tfrac{x}{|x|},t) = \sum_{\alpha=0}^{\infty}S_{n,\alpha}(|x|)Z_\alpha(x,t), \tag{6.2.5}$$

where the series converges absolutely, and uniformly on compact subsets of $\mathbb{B}$.

Proof. We first prove that the series (6.2.5) converges absolutely, and uniformly on compact subsets of $\mathbb{B}$. Consider $S_{n,\alpha}(r)$ given by

$$S_{n,\alpha}(r) = c_{n,\alpha}\sum_{k=0}^{\infty}\frac{(\alpha)_k(1-\frac{1}{2}n)_k}{(\alpha+\frac{1}{2}n)_k k!}r^{2k}.$$

Let $m = [n/2]$, and set $S_{n,\alpha}(r) = c_{n,\alpha}[P_m(r) + Q_m(r)]$, where

$$P_m(r) = \sum_{k=0}^{m-1}\frac{(\alpha)_k(1-\frac{1}{2}n)_k}{(\alpha+\frac{1}{2}n)_k k!}r^{2k};\ Q_m(r) = \sum_{k=m}^{\infty}\frac{(\alpha)_k(1-\frac{1}{2}n)_k}{(\alpha+\frac{1}{2}n)_k k!}r^{2k}. \tag{6.2.6}$$

If n is even, then $(1-\frac{1}{2}n)_k = 0$ for all $k \geq m$, and thus $Q_m(r) \equiv 0$. Now

$$|P_m(r)| \leq \sum_{k=0}^{m-1}\frac{(\alpha)_k|(1-\frac{1}{2}n)_k|}{(\alpha+\frac{1}{2}n)_k k!}r^{2k}.$$

Since $(\alpha)_k \big/ (\alpha+\frac{1}{2}n)_k \leq 1$ for all k,

$$|P_m(r)| \leq \sum_{k=0}^{m-1}\frac{|(1-\frac{1}{2}n)_k|}{k!} = C_n, \tag{6.2.7}$$

where C_n is a constant depending only on n.

Our next step is to obtain an estimate for $Q_m(r)$ when n is odd. For $k \geq m$ we have

$$(\gamma)_k = (\gamma)_m(\gamma + m)_{k-m}.$$

Thus

$$\begin{aligned} Q_m(r) &= \sum_{k=m}^{\infty} \frac{(\alpha)_k(1-\frac{1}{2}n)_k}{(\alpha+\frac{1}{2}n)_k k!} r^{2k} \\ &= \frac{(\alpha)_m(1-\frac{1}{2}n)_m}{(\alpha+\frac{1}{2}n)_m} r^{2m} \sum_{j=0}^{\infty} \frac{(\alpha+m)_j(1+m-\frac{1}{2}n)_j}{(\alpha+m+\frac{1}{2}n)_j(m+j)!} r^{2j}. \end{aligned} \tag{6.2.8}$$

As $(m+j)! \geq j!$ and $(1+m-\frac{1}{2}n) > 0$,

$$|Q_m(r)| \leq \frac{|(1-\frac{1}{2}n)_m|\Gamma(\alpha+m)\Gamma(\alpha+\frac{1}{2}n)}{\Gamma(\alpha)\Gamma(\alpha+\frac{1}{2}n+m)} F(\alpha+m, 1+m-\tfrac{1}{2}n; \alpha+m+\tfrac{1}{2}n; r^2).$$

But $F(\alpha+m, 1+m-\frac{1}{2}n; \alpha+m+\frac{1}{2}n; r^2)$ is an increasing function of r. Thus by identity (6.1.5)

$$F(\alpha+m, 1+m-\tfrac{1}{2}n; \alpha+m+\tfrac{1}{2}n; r^2) \leq \frac{\Gamma(\alpha+m+\frac{1}{2}n)\Gamma(n-1-m)}{\Gamma(\frac{1}{2}n)\Gamma(\alpha+n-1)}.$$

Therefore

$$|Q_m(r)| \leq \frac{|(1-\frac{1}{2}n)_m|\,\Gamma(n-1-m)}{\Gamma(\frac{1}{2}n)} \frac{\Gamma(\alpha+m)\Gamma(\alpha+\frac{1}{2}n)}{\Gamma(\alpha)\Gamma(\alpha+n-1)}. \tag{6.2.9}$$

But by (6.1.6)

$$c_{n,\alpha} = \frac{\Gamma(\frac{1}{2}n)\Gamma(\alpha+n-1)}{\Gamma(n-1)\Gamma(\alpha+\frac{1}{2}n)}.$$

Thus

$$|S_{n,\alpha}(r)| \leq C_n' \frac{\Gamma(\alpha+n-1)}{\Gamma(\alpha+\frac{1}{2}n)} + D_n \frac{\Gamma(\alpha+m)}{\Gamma(\alpha)},$$

where again C_n' and D_n are constants depending only on n. Using the fact that ([1, Identity 6.1.46])

$$\lim_{\alpha\to\infty} \alpha^{b-a} \frac{\Gamma(\alpha+a)}{\Gamma(\alpha+b)} = 1,$$

we have

$$\frac{\Gamma(\alpha+a)}{\Gamma(\alpha+b)} \approx \alpha^{a-b}. \tag{6.2.10}$$

Hence by the above

$$|Q_m(r)| \le D_n \alpha^{1/2} \quad \text{and} \quad |S_{n,\alpha}(r)| \le D_n' \alpha^{[n/2]}, \tag{6.2.11}$$

where C_n and D_n' are constants depending only on n. Also, for all $\zeta, t \in \mathbb{S}$,

$$|Z_\alpha(\zeta, t)| \le \|Z_\alpha\|_2^2 = h_\alpha,$$

where $h_\alpha = \dim(\mathcal{H}_\alpha(\mathbb{S}))$. By [10, Chapter 5]

$$h_\alpha = \binom{n+\alpha-1}{n-1} - \binom{n+\alpha-3}{n-1}.$$

But then $h_\alpha \le C\alpha^{n-2}$. Hence

$$\sum_{\alpha=0}^{\infty} |S_{n,\alpha}(|x|)|\,|Z_\alpha(x,\zeta)| \le C \sum_{\alpha=0}^{\infty} |x|^\alpha \alpha^p,$$

where $p = n + [n/2] - 2$. The series on the right however converges for all x, $|x| < 1$, and uniformly for $0 \le |x| \le \rho$, whenever $0 < \rho < 1$ is fixed. This proves our assertion.

It only remains to be shown that the series converges to $P_h(x,t)$. By (6.2.4), if $f \in L^2(\mathbb{S})$, then

$$f(\eta) = \sum_{\alpha=0}^{\infty} \langle f, Z_\eta^{(\alpha)} \rangle$$

in $L^2(\mathbb{S})$. In particular, by Theorem 6.1.1, for fixed $x \in \mathbb{B}$,

$$P_h(x,t) = \sum_{\alpha}^{\infty} P_h[Z_t^{(\alpha)}](x) = \sum_{\alpha=0}^{\infty} S_{n,\alpha}(|x|) Z_\alpha(x,t),$$

from which the result now follows. □

An immediate consequence of the previous theorem is the following corollary.

Corollary 6.2.3 *If $f \in L^2(\mathbb{S})$,*

$$P_h[f](x) = \sum_{\alpha=0}^{\infty} S_{n,\alpha}(|x|) \int_{\mathbb{S}} Z_\alpha(x,t) f(t)\, d\sigma(t),$$

where the series converges absolutely, and uniformly on compact subsets of $\mathbb{B}$.

Proof. For $f \in L^2(\mathbb{S})$, $f(\eta) = \sum_{\alpha=0}^{\infty} \langle f, Z_\eta^{(\alpha)} \rangle$ in $L^2(\mathbb{S})$. Thus

$$\int_{\mathbb{S}} P_h(x,\eta) f(\eta)\, d\sigma(\eta) = \sum_{\alpha=0}^{\infty} \int_{\mathbb{S}} P_h(x,\eta) \langle f, Z_\eta^{(\alpha)} \rangle \, d\sigma(\eta)$$
$$= \sum_{\alpha=0}^{\infty} \int_{\mathbb{S}} \int_{\mathbb{S}} P_h(x,\eta) f(t) Z_\alpha(\eta, t)\, d\sigma(t) d\sigma(\eta),$$

which by Fubini's theorem and Theorem 6.1.1

$$= \sum_{\alpha=0}^{\infty} S_{n,\alpha}(|x|) \int_{\mathbb{S}} f(t) Z_\alpha(x,t)\, d\sigma(t).$$

□

6.3 Spherical Harmonic Expansion of $\mathcal{H}$-Harmonic Functions

In this section we follow the methods of P. Ahern, J. Bruna, and C. Cascante [2] to obtain the following spherical harmonic expansion of $\mathcal{H}$-harmonic functions. The analogous result in [2] was proved for $\mathcal{M}$-harmonic functions on the unit ball of $\mathbb{C}^n$. The result has previously been proved by P. Jaming using the methods of [2] in his doctoral dissertation [41], and also earlier by K. Minemura in [57].

Theorem 6.3.1 *If u is an $\mathcal{H}$-harmonic function on $\mathbb{B}$, then*

$$u(r\zeta) = \sum_{\alpha} F(\alpha, 1 - \tfrac{1}{2}n; \alpha + \tfrac{1}{2}n; r^2) r^\alpha \varphi_\alpha(\zeta), \quad \zeta \in \mathbb{S},$$

where φ_α is a spherical harmonic of degree α. Moreover, the series converges absolutely and uniformly on every compact subset of $\mathbb{B}$.

Proof. For each r, $0 < r < 1$, by (6.2.4) the L^2-decomposition in harmonic polynomials of $u(r\zeta)$ is given by

$$u(r\zeta) = \sum_{\alpha=0}^{\infty} \int_{\mathbb{S}} Z_\alpha(\zeta,\eta) u(r\eta) d\sigma(\eta).$$

Let $\zeta \in \mathbb{S}$ and $\alpha \in \mathbb{N}$ be fixed, and for $r \in (-1, 1)$ let

$$f_\zeta(r) = \int_{\mathbb{S}} Z_\alpha(\zeta,\eta) u(r\eta) d\sigma(\eta).$$

Set

$$Lf = (1 - r^2)N^2 f + (n-2)(1 + r^2)Nf, \quad \text{where}$$
$$Nf = \langle x, \nabla f\rangle = r\frac{df}{dr}.$$

Also, let

$$\Delta_\sigma = \sum_{i<j} L_{i,j}^2 \quad \text{where} \quad L_{i,j} = x_i \frac{\partial}{\partial x_j} - x_j \frac{\partial}{\partial x_i}.$$

Since u is $\mathcal{H}$-harmonic, by Exercise 5.7.6

$$Lf_\zeta(r) = -(1 - r^2)\int_{\mathbb{S}} Z_\alpha(\zeta, \eta)\Delta_\sigma u(r\eta)d\sigma(\eta),$$

which by Exercises 6.4.3 and 6.4.4

$$= (1 - r^2)c_{\alpha,n} f_\zeta(r),$$

where $c_{\alpha,n} = \alpha[\alpha + (n-2)]$.

Since $f_\zeta(-r) = (-1)^\alpha f_\zeta(r)$, the function $f_\zeta(r)/r^\alpha$ is even on $(-1, 1)$. Hence, there exists $g(x)$ defined on $(0, 1)$ such that $f_\zeta(r) = r^\alpha g(r^2)$. Substituting into

$$Lf_\zeta(r) - c_{\alpha,n}(1 - r^2)f_\zeta(r) = 0$$

and simplifying gives

$$2r^2(1 - r^2)g''(r^2) + [(n + 2\alpha) + (n - 2\alpha - 4)r^2]g'(r^2) + \alpha(n-2)g(r^2) = 0,$$

which upon replacing r^2 by x and dividing by 2 yields the hypergeometric equation

$$x(1 - x)g''(x) + [(\alpha + \tfrac{n}{2}) - (\alpha - \tfrac{n}{2} + 2)x]g'(x) - \alpha(1 - \tfrac{n}{2})g(x) = 0.$$

With $a = \alpha$, $b = 1 - \frac{1}{2}n$, and $c = \alpha + \frac{1}{2}n$, this equation can be rewritten in standard form as

$$x(1 - x)g''(x) + \{c - (a + b + 1)x\}g'(x) - abg(x) = 0.$$

By Frobenius' theorem [30, Section 6.5], every solution of the above is a linear combination of two functions $g_1(x)$ and $g_2(x)$, whose behavior near $x = 0$ is respectively like 1 and x^{1-c}. Since $g(x)$ is bounded near 0, $g(x)$ is a multiple of the hypergeometric function (6.1.2). Therefore,

$$g(x) = c_\alpha(\zeta)F(\alpha, 1 - \tfrac{1}{2}n; \alpha + \tfrac{1}{2}n; x)$$

and

$$f_\zeta(r) = c_\alpha(\zeta)r^\alpha F(\alpha, 1 - \tfrac{1}{2}n; \alpha + \tfrac{1}{2}n; r^2).$$

For fixed r, as a function of $\zeta, f_\zeta(r) \in \mathcal{H}_\alpha(\mathbb{S})$. Hence there exists $\varphi_\alpha \in \mathcal{H}_\alpha(\mathbb{S})$ so that $c_\alpha(\zeta) = \varphi_\alpha(\zeta)$. Therefore

$$u(r\zeta) = \sum_\alpha F(\alpha, 1 - \tfrac{1}{2}n; \alpha + \tfrac{1}{2}n; r^2) r^\alpha \varphi_\alpha(\zeta). \tag{6.3.1}$$

It remains to be shown that the series (6.3.1) converges absolutely and uniformly on compact subsets of $\mathbb{B}$. For notational convenience, set $F_\alpha(r) = F(\alpha, 1 - \frac{1}{2}n; \alpha + \frac{1}{2}n; r^2)$. Also, as in Theorem 6.2.2 set $m = [n/2]$ and

$$P_m(\alpha, r) = \sum_{k=0}^{m-1} \frac{(\alpha)_k (1 - \frac{1}{2}n)_k}{(\alpha + \frac{1}{2}n)_k k!} r^{2k}; \quad Q_m(\alpha, r) = \sum_m^\infty \frac{(\alpha)_k (1 - \frac{1}{2}n)_k}{(\alpha + \frac{1}{2}n)_k k!} r^{2k}.$$

Let K be a compact subset of $\mathbb{B}$. Choose r_o, $0 < r_o < 1$, such that $K \subset B_{r_o}$. Then for $|\rho| < r_o$,

$$\sum_\alpha F_\alpha(\rho) \rho^\alpha \varphi_\alpha(\zeta) = \sum_\alpha \frac{F_\alpha(\rho)}{F_\alpha(r_o)} \left(\frac{\rho}{r_o}\right)^\alpha F_\alpha(r_o) r_o^\alpha \varphi_\alpha(\zeta).$$

But

$$\begin{aligned} |F_\alpha(r_o) r_o^\alpha \varphi_\alpha(\zeta)| &= \left| \int_{\mathbb{S}} Z_\alpha(\zeta, \eta) u(r_o \eta) d\sigma(\eta) \right| \\ &\le \|Z_\alpha(\zeta, \cdot)\|_2 \|u_{r_o}\|_2 \\ &\le h_\alpha^{\frac{1}{2}} \|u_{r_o}\|_2 \le C \alpha^{\frac{n}{2} - 1} \|u_{r_o}\|_2. \end{aligned}$$

Therefore

$$\sum_\alpha |F_\alpha(\rho) \rho^\alpha \varphi_\alpha(\zeta)| \le C \|u_{r_o}\|_2 \sum_\alpha \frac{|F_\alpha(\rho)|}{|F_\alpha(r_o)|} \left(\frac{|\rho|}{r_o}\right)^\alpha \alpha^{\frac{n}{2} - 1}.$$

Suppose n is even. Then $Q_m(\alpha, \rho) = 0$ for all ρ, and as in Theorem 6.2.2, $|P_m(\alpha, \rho)| \le C_n$. Also, since

$$(1 - m)_k = (-1)^k \frac{(m-1)!}{(m-k-1)!}$$

and $\displaystyle\lim_{\alpha \to \infty} \frac{(\alpha)_k}{(\alpha + \frac{1}{2}n)_k} = 1$, we have

$$\lim_{\alpha \to \infty} P(\alpha, r_o) = \sum_{k=0}^{m-1} (-1)^k \frac{(m-1)!}{(m-k-1)! k!} r_o^{2k} = (1 - r_o^2)^{m-1}.$$

Hence there exists an integer α_o such that $|P_m(\alpha, r_o)| \geq \frac{1}{2}(1 - r_o^2)^{m-1}$ for all $\alpha \geq \alpha_o$. Therefore

$$\sum_{\alpha=\alpha_o}^{\infty} |F_\alpha(\rho)\rho^\alpha \varphi_\alpha(\zeta)| \leq \frac{C\|u_{r_o}\|_2}{(1 - r_o^2)^{m-1}} \sum_{\alpha=\alpha_o}^{\infty} \left(\frac{|\rho|}{r_o}\right)^\alpha \alpha^{\frac{n}{2}-1}, \tag{6.3.2}$$

which converges for all ρ, $|\rho| < r_o$.

Suppose n is odd. Using the notation of [39, Theorem 7.25], set $\binom{\gamma}{0} = 1$, and for $k = 1, 2, \ldots,$

$$\binom{\gamma}{k} = \frac{\gamma(\gamma - 1)(\gamma - 2)\cdots(\gamma - k + 1)}{k!}.$$

Since $(1 - \frac{1}{2}n)_k = (-1)^k(\frac{1}{2}n - 1)(\frac{1}{2}n - 2)\cdots(\frac{1}{2}n - k)$, the series for $F_\alpha(r)$ can be expressed as

$$F_\alpha(r) = \sum_{k=0}^{\infty} \frac{(\alpha)_k}{(\alpha + \frac{1}{2}n)_k} \binom{\frac{1}{2}n - 1}{k} (-1)^k r^{2k}.$$

By Theorem 7.25 of [39], the series

$$\sum_{k=0}^{\infty} \left| \binom{\frac{1}{2}n - 1}{k} \right| \tag{6.3.3}$$

converges, and

$$\sum_{k=0}^{\infty} \binom{\frac{1}{2}n - 1}{k} (-1)^k r^{2k} = (1 - r^2)^{\frac{1}{2}n-1} \tag{6.3.4}$$

for all $r \in [0, 1]$, with the series converging uniformly and absolutely in $[0, 1]$.

For $N > \frac{1}{2}n$, let $P_N(\alpha, r)$ and $Q_N(\alpha, r)$ be as defined in (6.2.6). Then

$$|F_\alpha(r_o)| \geq |P_N(\alpha, r_o)| - |Q_N(\alpha, r_o)|.$$

Since $(\alpha)_k < (\alpha + \frac{1}{2}n)_k$,

$$|Q_N(\alpha, r)| \leq \sum_{k=N}^{\infty} \left| \binom{\frac{1}{2}n - 1}{k} \right|,$$

and since the series converges, given $\epsilon > 0$, there exists an integer N_1 such that $|Q_N(\alpha, r)| < \epsilon$ for all $N \geq N_1$ and all non-negative integers α and $r \in [0, 1]$. Also, since

$$\lim_{N\to\infty} |P_N(r_o)| = (1 - r_o^2)^{\frac{1}{2}n-1},$$

there exists an integer N_2 such that $|P_N(r_o)| \geq (1-r_o^2)^{\frac{1}{2}n-1} - \epsilon$ for all $N \geq N_2$. Fix an $N \geq \max\{N_1, N_2\}$. For this N, since

$$\lim_{\alpha\to\infty} |P_N(\alpha, r_o)| = |P_N(r_o)|,$$

there exists an integer α_o such that $|P_N(\alpha, r_o)| > |P_N(r_o)| - \epsilon$ for all $\alpha \geq \alpha_o$. Take $\epsilon = \frac{1}{6}(1-r_o^2)^{\frac{1}{2}n-1}$. Then for this epsilon, there exists an α_o such that

$$|F_\alpha(r_o)| > \tfrac{1}{2}(1-r_o^2)^{\frac{1}{2}n-1}$$

for all $\alpha \geq \alpha_o$. Finally, as a consequence of (6.2.7) and (6.2.11), we have that $|F_\alpha(\rho)| \leq C_n \alpha^{\frac{1}{2}}$ for all $\rho \in (-1, 1)$. Hence, as in (6.3.2), we have that the series converges absolutely and uniformly for all ρ, $|\rho| < r_o$. □

6.4 Exercises

6.4.1. (a) Find the zonal harmonic expansion of the Euclidean Poisson kernel.
(b) Let Q be a polynomial on $\mathbb{R}^n$ of degree m. Prove that the Euclidean Poisson integral $P_e[Q]$ of Q is a polynomial of degree at most m, and that

$$P_e[Q](x) = \sum_{k=0}^{m} \int_{\mathbb{S}} Z_k(x,t)Q(t)d\sigma(t).$$

6.4.2. Let Q be a polynomial on $\mathbb{S}$, and let $P_h[Q]$ denote the Poisson integral of Q. Prove that $|\nabla P_h[Q](x)|$ is continuous on $\overline{\mathbb{B}}$.

6.4.3. As in Exercise 3.5.6 let

$$\Delta_\sigma = \sum_{i<j} L_{i,j}^2 \quad \text{where} \quad L_{i,j} = x_i\frac{\partial}{\partial x_j} - x_j\frac{\partial}{\partial x_i}.$$

(a) Prove that for $f \in C^2(\mathbb{B})$

$$\Delta_\sigma f(x) = |x|^2 \Delta f(x) - (n-2)\langle x, \nabla f(x)\rangle - \langle x, \nabla\langle x, \nabla f(x)\rangle\rangle.$$

(b) With Z_α, $\alpha = 0, 1, 2, \ldots$ as given in Theorem 6.2.1, prove that

$$\Delta_\sigma Z_\alpha(x,\eta) = -\alpha[\alpha + (n-2)]Z_\alpha(x,\eta).$$

6.4.4. (a) If $u_\alpha(x) = \int_{\mathbb{S}} Z_\alpha(x,\zeta)u(\zeta)d\sigma(\zeta)$, prove that $\Delta_\sigma u_\alpha(x) = -c_{n,\alpha}u_\alpha(x)$ where $c_{n,\alpha} = \alpha[\alpha + (n-2)]$.
(b) If $u, v \in L^2(\mathbb{S})$, prove that $\langle u, \Delta_\sigma v\rangle = \langle \Delta_\sigma u, v\rangle$.

6.4.5. Prove that for $c > b > 0$,

$$F(a,b;c;r) = \frac{\Gamma(c)}{\Gamma(b)\Gamma(c-b)} \int_0^1 t^{b-1}(1-t)^{c-b-1}(1-rt)^{-a}dt.$$

6.4.6. Prove the following result of H. Samii [73]. If u is $\mathcal{H}$-harmonic on $\mathbb{B}$, then there exists a Euclidean harmonic function v with $v(0) = 0$ such that

$$u(x) = u(0) + \int_0^1 v(tx)[(1-t)(1-tr^2)]^{n/2-1}\frac{dt}{t}.$$

7

Hardy-Type Spaces of $\mathcal{H}$-Subharmonic Functions

In this chapter we consider Hardy-type $\mathcal{H}^p$ spaces of both $\mathcal{H}$-harmonic and $\mathcal{H}$-subharmonic functions on $\mathbb{B}$. The main result of Section 7.1 is a Poisson integral formula for $\mathcal{H}$-harmonic functions in the space $\mathcal{H}^p$, $1 \leq p \leq \infty$. At the same time we also prove the existence of an $\mathcal{H}$-harmonic majorant for an $\mathcal{H}$-subharmonic function satisfying an H^p growth condition. In Section 7.3 we take up the general question of the existence of $\mathcal{H}$-harmonic majorants for $\mathcal{H}$-subharmonic functions as well as the existence of a least $\mathcal{H}$-harmonic majorant. In Section 7.4 we consider a generalization of a theorem of L. Garding and L. Hörmander [26] concerning Hardy–Orlicz spaces of $\mathcal{H}$-subharmonic functions.

We begin the chapter with the following definitions.

Definition 7.0.1 *For* $0 < p \leq \infty$, *we denote by* $\mathcal{S}^p$ *the* **Hardy-type** *space of non-negative continuous* $\mathcal{H}$*-subharmonic functions* f *on* $\mathbb{B}$ *for which*

$$\|f\|_p^p = \sup_{0<r<1} \int_{\mathbb{S}} f^p(rt)\,d\sigma(t) < \infty. \tag{7.0.1}$$

When $p = \infty$, *we set* $\|f\|_\infty = \sup_{x\in\mathbb{B}} f(x)$.

In the case $0 < p < 1$, as a general rule we will only be interested in non-negative $\mathcal{H}$-subharmonic functions for which f^p is also $\mathcal{H}$-subharmonic. In the case that f^p is $\mathcal{H}$-subharmonic, then by Theorem 5.4.2

$$\|f\|_p^p = \lim_{r\to 1} \int_{\mathbb{S}} f^p(rt)\,d\sigma(t).$$

Likewise, for $0 < p < \infty$, we let

$$\mathcal{H}^p = \{f : f \text{ is } \mathcal{H}\text{-harmonic with } |f| \in \mathcal{S}^p\}.$$

The space $\mathcal{H}^p$ is called the **Hardy space** of $\mathcal{H}$-harmonic functions on $\mathbb{B}$. Also, for $\lambda \ge 0$ and $0 < p < \infty$ we let

$$\mathcal{H}^p_\lambda = \{f \in \mathcal{H}_\lambda : |f| \in \mathcal{S}^p\}.$$

By Exercise 5.7.4, if $f \in \mathcal{H}_\lambda$, $\lambda \ge 0$, then $|f|$ is $\mathcal{H}$-subharmonic on $\mathbb{B}$.

As a general rule, in discussions of $\mathcal{H}^p$ or $\mathcal{H}^p_\lambda$ we will primarily be interested in the case $p \ge 1$. However, the definitions still make sense for $0 < p < 1$. In fact, if f is a positive $\mathcal{H}$-harmonic function on $\mathbb{B}$, then for $0 < p < 1$, f^p is $\mathcal{H}$-superharmonic and thus $f \in \mathcal{H}^p$ for all p, $0 < p \le 1$.

7.1 A Poisson Integral Formula for Functions in $\mathcal{H}^p$, $1 \le p \le \infty$

We now prove the Poisson integral formula for $\mathcal{H}$-harmonic functions in $\mathcal{H}^p$, $1 \le p \le \infty$, while at the same time proving the existence of an $\mathcal{H}$-harmonic majorant for an $\mathcal{H}$-subharmonic function f in $\mathcal{S}^p$, that is, $f(x) \le H(x)$ for some $\mathcal{H}$-harmonic function H.

Theorem 7.1.1 *Let f be a continuous non-negative $\mathcal{H}$-subharmonic function on $\mathbb{B}$ with $f \in \mathcal{S}^p$ for some p, $1 \le p \le \infty$.*

(a) *If $1 < p \le \infty$, then there exists $\hat{f} \in L^p(\mathbb{S})$ such that*

$$f(x) \le P_h[\hat{f}](x) \quad \text{for all } x \in \mathbb{B} \tag{7.1.1}$$

with $\|f\|_p = \|\hat{f}\|_p$.

(b) *If $p = 1$, then there exists a Borel measure ν_f on $\mathbb{B}$ such that*

$$f(x) \le P_h[\nu_f](x) \quad \text{for all } x \in \mathbb{B}. \tag{7.1.2}$$

(c) *If f is $\mathcal{H}$-harmonic on $\mathbb{B}$ with $|f| \in \mathcal{S}^p$, that is, $f \in \mathcal{H}^p$, for some p, $1 \le p \le \infty$, then equality holds in (7.1.1) and (7.1.2). Furthermore, in this case the measure ν_f is a signed Borel measure on $\mathbb{S}$.*

Since non-negative $\mathcal{H}$-harmonic functions satisfy (7.0.1) with $p = 1$, we have the following corollary.

Corollary 7.1.2 *If h is a non-negative $\mathcal{H}$-harmonic function on $\mathbb{B}$, then there exists a Borel measure μ on $\mathbb{S}$ such that $h(x) = P_h[\mu](x)$ for all $x \in \mathbb{B}$.*

Remarks 7.1.3 **(1)** *For $\mathcal{H}$-harmonic functions f, the function $\hat{f}$ and measure ν_f are called the* **boundary function** *and* **boundary measure** *of f. Furthermore, if*

$$d\nu_f = \hat{f}\, d\sigma + \nu_s$$

is the Lebesgue decomposition of f (see [71, p. 278]) where $\hat{f} \in L^1(\mathbb{S})$ and ν_s is singular with respect to σ, then as we will prove in Theorem 8.3.3, for every $\alpha > 1$,

$$\lim_{\substack{x \to \zeta \\ x \in \Gamma_\alpha(\zeta)}} f(x) = \hat{f}(\zeta) \quad \sigma\text{ - a.e. on } \mathbb{S}, \tag{7.1.3}$$

where $\Gamma_\alpha(\zeta) = \{x \in \mathbb{B} : |x - \zeta| < \alpha(1 - |x|)\}$.

(2) *For invariant harmonic functions, the case $p = \infty$ was originally proved by H. Furstenberg ([24]) for symmetric spaces of noncompact type. The L^p statements were proved by A. Koranyi in [46] for $\mathcal{M}$-harmonic functions in the unit ball of $\mathbb{C}^m$ and by the author [78] for symmetric spaces of noncompact type. In the setting of rank one symmetric spaces, $\mathcal{H}^p$ spaces have also been considered by P. Cifuentis in [14], [15]. The proof given here is a variation of an elementary proof using an equicontinuity argument which for $\mathcal{M}$-harmonic functions is due to D. Ullrich [93]. (See also [72, Theorem 4.3.3] and [84, Theorem 5.8].) The result for Euclidean subharmonic functions on the unit ball in $\mathbb{R}^n$ is due to L. Garding and L. Hörmander [26]. An alternate proof of Theorem 7.1.1 for $\mathcal{H}$-harmonic functions was given by P. Jaming in [42, Proposition 2]. In [31] the $\mathcal{H}^p$ spaces were defined for all p, $0 < p < \infty$, in terms of the radial maximal function (see Definition 8.2.1).*

Proof. Let $h : O(n) \to [0, \infty)$ be a continuous function satisfying

$$\int_{O(n)} h(A)dA = 1,$$

where dA is the normalized Haar measure on $O(n)$. Let $f \in \mathcal{S}^p$ ($\mathcal{H}^p$) for some p, $1 \le p \le \infty$. Define the function G on $\mathbb{B}$ by

$$G(x) = \int_{O(n)} f(Ax)h(A)dA. \tag{7.1.4}$$

Consider

$$\int_{\mathbb{S}} G(\varphi_a(rt))d\sigma(t) = \int_{\mathbb{S}} \int_{O(n)} f(A\varphi_a(rt))h(A)dA d\sigma(t),$$

which by Fubini's theorem

$$= \int_{O(n)} \left[\int_{\mathbb{S}} f(A\varphi_a(rt))d\sigma(t) \right] h(A)dA.$$

By Exercise 2.4.5, $A\varphi_a = \varphi_{Aa} \circ B$ for some $B \in O(n)$. Therefore,

$$\int_{\mathbb{S}} f(A\varphi_a(rt))d\sigma(t) = \int_{\mathbb{S}} f(\varphi_{Aa}(rBt))d\sigma(t),$$

which since σ is invariant under $O(n)$

$$= \int_{\mathbb{S}} f(\varphi_{Aa}(rt))d\sigma(t) \ge f(Aa),$$

with equality if f is $\mathcal{H}$-harmonic. Therefore

$$\int_{\mathbb{S}} G(\varphi_a(rt))d\sigma(t) \ge \int_{O(n)} f(Aa)h(A)dA = G(a),$$

again with equality if f is $\mathcal{H}$-harmonic. Thus G is $\mathcal{H}$-subharmonic ($\mathcal{H}$-harmonic).

Since $\|f\|_p < \infty$, by Hölder's inequality applied to (7.1.4) for $x = r\zeta$, $\zeta \in \mathbb{S}$,

$$|G(x)| \le \left[\int_{O(n)} |f(Ar\zeta)|^p dA\right]^{1/p} \|h\|_q \le \|f\|_p \|h\|_q$$

where $\frac{1}{p} + \frac{1}{q} = 1$. In the above we have used the fact that

$$\int_{O(n)} |f(Ar\zeta)|^p dA = \int_{\mathbb{S}} \int_{O(n)} |f(Ar\zeta)|^p dA d\sigma(\zeta),$$

which by (3.3.3)

$$= \int_{\mathbb{S}} |f(r\zeta)|^p d\sigma(\zeta).$$

Also, by the continuous version of Minkowski's inequality ([18, VI.11.13]), for $1 \le p < \infty$,

$$\begin{aligned}\left[\int_{\mathbb{S}} |G(r\zeta)|^p d\sigma(\zeta)\right]^{1/p} &\le \int_{O(n)} h(A) \left[\int_{\mathbb{S}} |f(Ar\zeta)|^p d\sigma(\zeta)\right]^{1/p} dA \\ &\le \|f\|_p. \end{aligned} \tag{7.1.5}$$

Clearly the result is also true for $p = \infty$.

We now show that the family $\{G_r : 0 < r < 1\}$, where $G_r(t) = G(rt)$, is an **equicontinuous** family on $\mathbb{S}$; that is, to each $\epsilon > 0$ there exists a $\delta > 0$ such that

$$|G_r(x) - G_r(y)| < \epsilon$$

for all $r, 0 < r < 1$, and all $x, y \in \mathbb{S}$ with $|x - y| < \delta$. Let $\epsilon > 0$ be fixed. Since h is continuous on $O(n)$, there exists a neighborhood $\mathcal{V}$ of the identity I such that

$$|h(A) - h(AB^{-1})| < \epsilon$$

for all $A \in O(n)$, $B \in \mathcal{V}$. Furthermore, since

$$G_r(Bx) = \int_{O(n)} f(rABx)h(A)dA = \int_{O(n)} f(rAx)h(AB^{-1})dA,$$

we have

$$\begin{aligned}|G_r(x) - G_r(Bx)| &\leq \int_{O(n)} |f(rAx)||h(A) - h(AB^{-1})|dA \\ &\leq \epsilon \|f\|_1 \leq \epsilon \|f\|_p\end{aligned}$$

for all $B \in \mathcal{V}$. The mapping $B \to Bx$ is a one-to-one mapping of $O(n)$ onto $\mathbb{S}$. Thus there exists a $\delta > 0$ such that $|x - y| < \delta$ implies $y = Bx$ for some $B \in \mathcal{V}$. Therefore

$$|G_r(x) - G_r(y)| < \epsilon \|f\|_p$$

for all $x, y \in \mathbb{S}$ with $|x - y| < \delta$ and all r, $0 < r < 1$. Hence $\{G_r\}$ is equicontinuous on $\mathbb{S}$. Hence by the Ascoli–Arzelá theorem [71], there exists a sequence $r_k \to 1$ such that G_{r_k} converges uniformly to a continuous function g on S.

Let

$$\epsilon_k = \sup_{t \in \mathbb{S}} |G(r_k t) - P_h[g](r_k t)|.$$

Since $G_{r_k} \to g$ uniformly on $\mathbb{S}$, and since g is continuous, from the proof of Theorem 5.3.3 we also have that $P_h[g](r_k t) \to g(t)$ uniformly. Thus $\epsilon_k \to 0$ as $k \to \infty$. Hence by the maximum principle

$$G(x) \leq P_h[g](x) + \epsilon_k \quad \text{for all } x,\ |x| \leq r_k.$$

If f is $\mathcal{H}$-harmonic, then

$$|G(x) - P_h[g](x)| \leq \epsilon_k \quad \text{for all } x,\ |x| \leq r_k.$$

Letting $k \to \infty$ we have

$$G(x) \leq P_h[g](x) \tag{7.1.6}$$

for all $x \in \mathbb{B}$ with equality if f is $\mathcal{H}$-harmonic. Since

$$\int_{\mathbb{S}} |G(r_k \zeta)|^p d\sigma(\zeta) \leq \|f\|_p^p$$

and $G_{r_k} \to g$ uniformly on $\mathbb{S}$, we have

$$\|g\|_p \leq \|f\|_p, \quad 1 \leq p \leq \infty.$$

To conclude the proof, let $\{h_j\}$ be a sequence of continuous functions on $O(n)$ which forms an approximate identity, that is, $h_j \geq 0$ for all j,

$\int_{O(n)} h_j dA = 1$, and $\lim_{j\to\infty} \int_{O(n)\setminus\mathcal{V}} h_j dA = 0$ for every neighborhood $\mathcal{V}$ of the identity. For each j let G_j be as defined by (7.1.4). Since f is continuous, as in the proof of Theorem 5.3.3,

$$\lim_{j\to\infty} G_j(x) = f(x) \quad \text{for all } x \in \mathbb{B}. \tag{7.1.7}$$

By the above, each $G_j = P_h[g_j]$ for some $g_j \in L^p(\mathbb{S})$ with $\|g_j\|_p \le \|f\|_p$. If $p > 1$, then by the Alaoglu theorem (see [71, p. 237]), some subsequence of $\{g_j\}$, which we denote by $\{g_j\}$, converges in the weak* topology of $L^p(\mathbb{S})$ to some $\hat{f} \in L^p(\mathbb{S})$, that is, $\int g_j h d\sigma \to \int \hat{f} h d\sigma$ for every $h \in L^q(\mathbb{S})$, $\frac{1}{p} + \frac{1}{q} = 1$. In particular $\lim_{j\to\infty} P_h[g_j] = P_h[\hat{f}]$. Thus by (7.1.6) and (7.1.7),

$$f(x) \le P_h[\hat{f}](x)$$

with equality if f is $\mathcal{H}$-harmonic. Furthermore, since $\|g_j\|_p \le \|f\|_p$ we have

$$\|\hat{f}\|_p \le \|f\|_p.$$

Also, since $f(x) \le P_h[\hat{f}](x)$ (or $f(x) = P_h[\hat{f}](x)$ for $\mathcal{H}$-harmonic f) we also have $\|f\|_p \le \|\hat{f}\|_p$.

For the case $p = 1$, some subsequence of $\{g_j\}$ converges weak* in the dual of $C(\mathbb{S})$. Thus $f(x) \le P_h[\nu_f](x)$ for some Borel measure ν_f on $\mathbb{S}$. If f is $\mathcal{H}$-harmonic, then $f(x) = P_h[\nu_f](x)$ for some signed Borel measure ν_f on $\mathbb{S}$. □

7.2 Completeness of $\mathcal{H}^p$, $0 < p \le \infty$

For $0 < p \le \infty$, $f, g \in \mathcal{H}^p$, define

$$d_p(f,g) = \begin{cases} \|f-g\|_p, & p \ge 1, \\ \|f-g\|_p^p, & 0 < p < 1. \end{cases} \tag{7.2.1}$$

Then d_p is a metric on $\mathcal{H}^p$. This follows from the fact that for $p \ge 1$, $\|\cdot\|_p$ is a norm on $\mathcal{H}^p$, while for $0 < p < 1$, $\|f\|_p^p$ is a p-norm. If X is a vector space (of functions), a function $\|\cdot\| : X \to [0,\infty)$ is a **norm** on X if (i) $\|f\| = 0$ if and only if $f = 0$, (ii) $\|f+g\| \le \|f\| + \|g\|$ for all $f,\ g \in X$, and (iii) $\|af\| = |a|\,\|f\|$ for all $a \in \mathbb{R}$ and $f \in X$. For $0 < p < 1$, $\|\cdot\|$ is a p-**norm** if $\|\cdot\|^p$ satisfies (i) and (ii) above, and in addition (iii*) $\|af\|^p = |a|^p\|f\|^p$ for all $a \in \mathbb{R}$, $f \in X$.

In this section we prove that the metric space $(\mathcal{H}^p, d_p)$ is complete for all p, $0 < p < \infty$. The case $p = \infty$ is left to the exercises. We begin with the following lemma.

Lemma 7.2.1 *If $f \in \mathcal{S}^p$, $0 < p < \infty$, then there exists a constant C, independent of f, such that*

$$f^p(x) \le \frac{C}{(1-|x|^2)^{n-1}} \|f\|_p^p$$

for all $x \in \mathbb{B}$.

Proof. For $1 \le p < \infty$, the result is an immediate consequence of the Poisson integral formula for functions in $\mathcal{S}^p$ (see Exercise 7.5.3).

Fix δ, $0 < \delta < \frac{1}{2}$. Since f is $\mathcal{H}$-subharmonic, by Theorem 4.7.3 there exists a constant C_δ such that

$$f^p(a) \le C_{\delta,p} \int_{E(a,\delta)} f^p(x) d\tau(x).$$

By Exercise 2.4.1(b), if $x \in E(a,\delta)$, then

$$c_\delta(1-|a|^2) \le (1-|x|^2) \le c_\delta^{-1}(1-|a|^2),$$

where $c_\delta = (1-\delta)/(1+\delta)$. Let

$$A_\delta(a) = \{r\zeta : \zeta \in \mathbb{S},\, c_\delta(1-|a|^2) \le (1-r^2) \le c_\delta^{-1}(1-|a|^2)\}.$$

Thus

$$\begin{aligned} f^p(a) &\le C_{\delta,p} \int_{A_\delta(a)} f^p(x) d\tau(x) \\ &\le \frac{C\|f\|_p^p}{(1-|a|^2)^n} \int_{\sqrt{1-c_\delta^{-1}(1-|a|^2)}}^{\sqrt{1-c_\delta(1-|a|^2)}} r^{n-1} dr \\ &\le \frac{C\|f\|_p^p}{(1-|a|^2)^{n-1}}. \end{aligned}$$

□

Theorem 7.2.2 *The metric space* $(\mathcal{H}^p, d_p)$ *is complete for all* p, $0 < p < \infty$.

Proof. The case $p = \infty$ is left as an exercise (Exercise 7.5.4). Suppose $\{f_n\}$ is a Cauchy sequence in $(\mathcal{H}^p, d_p)$ and K is a compact subset of $\mathbb{B}$. By Lemma 7.2.1, there exists a constant C_K such that

$$|f_n(x) - f_m(x)| \le C_K \|f_n - f_m\|_p^p$$

for all $n, m \in \mathbb{N}$ and all $x \in K$. Thus $\{f_n\}$ is a uniform Cauchy sequence on each compact subset of $\mathbb{B}$ and as a consequence converges uniformly on compact sets to a function f. Let

$$f(x) = \lim_{m\to\infty} f_m(x).$$

Since the convergence is uniform on compact sets, f is $\mathcal{H}$-harmonic on $\mathbb{B}$. Furthermore, by Fatou's lemma, $f \in \mathcal{H}^p$. It remains to be shown that $f_n \to f$

in $\|\cdot\|_p$. Let $\epsilon > 0$ be given. Since $\{f_n\}$ is Cauchy, there exists an integer n_o such that $\|f_n - f_m\|_p^p < \epsilon$ for all $n, m \geq n_o$. But for each $r \in (0, 1)$,

$$\begin{aligned}\int_{\mathbb{S}} |f(rt) - f_m(rt)|^p d\sigma(t) &\leq \varliminf_{n\to\infty} \int_{\mathbb{S}} |f_n(rt) - f_m(rt)|^p d\sigma(t) \\ &\leq \varliminf_{n\to\infty} \|f_n - f_m\|_p^p \leq \epsilon.\end{aligned}$$

Taking the supremum over $r \in (0, 1)$ gives $\|f - f_m\|_p^p \leq \epsilon$ for all $m \geq n_o$; that is, $\{f_n\}$ converges to f in the metric of $\mathcal{H}^p$. □

7.3 $\mathcal{H}$-Harmonic Majorants for $\mathcal{H}$-Subharmonic Functions

In Theorem 7.1.1 we proved that if f is a continuous non-negative $\mathcal{H}$-subharmonic function satisfying

$$\lim_{r\to 1} \int_{\mathbb{S}} f(rt) d\sigma(t) < \infty,$$

then f has an $\mathcal{H}$-harmonic majorant of the form $P_h[\nu]$ where ν is a Borel measure on $\mathbb{S}$. In this section we consider the existence of $\mathcal{H}$-harmonic majorants for general $\mathcal{H}$-subharmonic functions on $\mathbb{B}$.

Definition 7.3.1 *An $\mathcal{H}$-subharmonic function f on $\mathbb{B}$ has an* **$\mathcal{H}$-harmonic majorant** *if there exists an $\mathcal{H}$-harmonic function h such that $f(x) \leq h(x)$ for all $x \in \mathbb{B}$. Furthermore, if there exists an $\mathcal{H}$-harmonic function F satisfying*

(a) *$f(x) \leq F(x)$ for all $x \in \mathbb{B}$, and*

(b) *$F(x) \leq G(x)$ for any $\mathcal{H}$-harmonic majorant G of f,*

then F is called the **least $\mathcal{H}$-harmonic majorant** *of f and will be denoted by F_f.*

The following theorem provides necessary and sufficient conditions for the existence of a least $\mathcal{H}$-harmonic majorant.

Theorem 7.3.2 *Let f be $\mathcal{H}$-subharmonic on $\mathbb{B}$. Then the following are equivalent:*

(a) *f has a least $\mathcal{H}$-harmonic majorant on $\mathbb{B}$.*

(b) *f has an $\mathcal{H}$-harmonic majorant on $\mathbb{B}$.*

(c) $\displaystyle\lim_{r\to 1} \int_{\mathbb{S}} f(rt) d\sigma(t) < \infty.$

For the proof of the theorem we will need the following generalization of Harnack's inequality for $\mathcal{H}$-harmonic functions.

Lemma 7.3.3 (Harnack's Inequality) *Let Ω be an open-connected subset of $\mathbb{B}$ and $a \in \Omega$. If K is a compact subset of Ω, then there exists a constant C_K such that*

$$h(x) \leq C_K h(a)$$

for all $x \in K$ and all non-negative $\mathcal{H}$-harmonic functions on Ω.

Proof. Let $y \in \Omega$ be arbitrary and $\delta > 0$ such that $\overline{E(y, 4\delta)} \subset \Omega$. If $y_1, y_2 \in E(y, \delta)$, then

$$E(y_1, \delta) \subset E(y, 2\delta) \subset E(y_2, 3\delta).$$

Therefore, by (4.3.2),

$$h(y_1) = \frac{1}{\tau(B_\delta)} \int_{E(y_1,\delta)} h(x)d\tau(x),$$

which since h is non-negative

$$\begin{aligned} &\leq \frac{1}{\tau(B_\delta)} \int_{E(y_2,3\delta)} h(x)d\tau(x) \\ &= \frac{\tau(B_{3\delta})}{\tau(B_\delta)} h(y_2) = C_\delta h(y_2). \end{aligned}$$

Choose $\delta > 0$ such that $\text{dist}_h(K, \partial\Omega) > 4\delta$ where for sets A and B

$$\text{dist}_h(A, B) = \inf\{d_h(x, y) : x \in A,\ y \in B\}.$$

Since K is compact and $\partial\Omega$ is closed, one can easily show that $\text{dist}_h(K, \partial\Omega) > 0$. Since K is compact, K can be covered by a finite number of hyperbolic balls $E(y_i, \delta)$, $i = 1, \ldots, N$. Hence for any $y \in K$, $h(y) \leq C_\delta h(y_i)$ for some $i \in \{1, \ldots, N\}$.

Since Ω is connected, for each i choose an arc $A_i \subset \Omega$ joining a to y_i and $\delta_i > 0$ such that $\text{dist}_h(A_i, \partial\Omega) > 4\delta_i$. Each arc A_i can be covered by a finite number N_i of balls $E(x_j, \delta_i)$, $j = 1, \ldots, N_i$. Hence

$$h(y_i) \leq C_{\delta_i}^{N_i} h(a).$$

If we let $C_K = \max\{C_\delta C_{\delta_i}^{N_i} : i = 1, \ldots, N\}$, then for any $y \in K$,

$$h(y) \leq C_K h(a).$$

□

Lemma 7.3.4 *Let f be $\mathcal{H}$-subharmonic on $\mathbb{B}$. Then for each r, $0 < r < 1$, there exists an $\mathcal{H}$-harmonic function F^r on B_r such that*

(a) *$f(x) \leq F^r(x)$ for all $x \in B_r$, and*

(b) $\displaystyle\int_{\mathbb{S}} f(rt)d\sigma(t) = F^r(0).$

Furthermore, if F is $\mathcal{H}$-harmonic on an open subset Ω of $\mathbb{B}$ with $\overline{B_r} \subset \Omega$ and $F(x) \geq f(x)$ for all $x \in \Omega$, then

(c) *$F^r(x) \leq F(x)$ for all $x \in B_r$.*

(d) *In particular, if $0 < r < \rho < 1$, then $F^r(x) \leq F^\rho(x)$ for all $x \in B_r$.*

Proof. Fix r, $0 < r < 1$. Since f is upper semicontinuous on $S_r = r\mathbb{S}$, there exists a decreasing sequence $\{f_n\}$ of continuous functions on S_r such that $\lim_{n\to\infty} f_n(rt) = f(rt)$ for all $t \in \mathbb{S}$. For each n, by Theorem 5.4.1 there exists an $\mathcal{H}$-harmonic function H_n on B_r with $H_n(rt) = f_n(rt)$. By Harnack's inequality, for each compact subset K of B_r there exists a constant C_K such that for $m \geq n$ we have

$$0 \leq H_m(x) - H_n(x) \leq C_K[H_m(0) - H_n(0)] = C_K \int_{\mathbb{S}} [f_m(rt) - f_n(rt)]\, d\sigma(t).$$

Therefore,

$$H^r(x) = \lim_{n\to\infty} H_n(x)$$

exists for all $x \in B_r$ and is $\mathcal{H}$-harmonic on B_r with

$$H^r(0) = \lim_{n\to\infty} \int_{\mathbb{S}} f_n(rt) d\sigma(t) = \int_{\mathbb{S}} f(rt) d\sigma(t). \tag{7.3.1}$$

Suppose F is $\mathcal{H}$-harmonic on a domain $\Omega \supset \overline{B_r}$ with $F(x) \geq f(x)$ for all $x \in \Omega$. With $\{f_n\}$ as above, set

$$g_n(rt) = \min\{f_n(rt), F(rt)\},$$

and let G_n be the corresponding $\mathcal{H}$-harmonic function on B_r. Since $g_n \leq f_n$ on S_r we have $G_n(x) \leq F_n(x)$ for all $x \in B_r$. But $\{g_n\}$ is a non-increasing sequence of functions on S_r with $\lim g_n(rt) = f(rt)$ for all $t \in \mathbb{S}$. Thus by (7.3.1),

$$\lim_{n\to\infty} G_n(0) = \int_{\mathbb{S}} f(rt)\, d\sigma(t) = \lim_{n\to\infty} F_n(0).$$

Therefore as a consequence of Harnack's inequality,

$$F^r(x) = \lim_{n\to\infty} F_n(x) = \lim_{n\to\infty} G_n(x).$$

However, by the maximum principle, $G_n(x) \leq F(x)$ for all n and all $x \in B_r$. Thus $F^r(x) \leq F(x)$ for all $x \in B_r$. The result (d) is an immediate consequence of (c). □

Proof of Theorem 7.3.2. Clearly (a) ⇒ (b) ⇒ (c). Suppose that (c) holds. Choose an increasing sequence $\{r_n\}$ with $r_n \to 1$. For each n, let $F^{(n)}$ be the $\mathcal{H}$-harmonic function on B_{r_n} satisfying the conclusion of Lemma 7.3.4. By part (a) of the lemma we have

$$F^{(n)}(x) \leq F^{(n+1)}(x) \quad \text{for all } x, |x| < r_n.$$

Since

$$\lim_{n\to\infty} F_n(0) = \lim_{n\to\infty} \int_{\mathbb{S}} f(r_n t)d\sigma(t) < \infty,$$

by Harnack's theorem (Exercise 7.5.1),

$$F_f(x) = \lim_{n\to\infty} F^{(n)}(x) \tag{7.3.2}$$

is $\mathcal{H}$-harmonic on $\mathbb{B}$ and satisfies $F_f(x) \geq f(x)$ for all $x \in \mathbb{B}$. Suppose G is an $\mathcal{H}$-harmonic majorant of f. Then by part (c) of the lemma $G(x) \geq F^{(n)}(x)$ for all $x \in B_{r_n}$. Therefore $G(x) \geq F_f(x)$ for all $x \in \mathbb{B}$, and thus F_f is the least $\mathcal{H}$-harmonic majorant of f. □

Corollary 7.3.5 *Let $f \leq 0$ be $\mathcal{H}$-subharmonic on $\mathbb{B}$. Then the least $\mathcal{H}$-harmonic majorant of f is the zero function if and only if*

$$\lim_{r\to 1} \int_{\mathbb{S}} f(rt)\, d\sigma(t) = 0.$$

If F_f is the least $\mathcal{H}$-harmonic majorant of f given by (7.3.2), then since

$$F_f(0) = \lim_{r\to 1} \int_{\mathbb{S}} f(rt)d\sigma(t)$$

we have

$$\lim_{r\to 1} \int_{\mathbb{S}} | F_f(rt) - f(rt)| d\sigma(t) = 0. \tag{7.3.3}$$

For a function f on $\mathbb{B}$, let

$$f^+(x) = \max\{f(x), 0\}.$$

The function f^+ is called the **positive part** of f. If f is $\mathcal{H}$-subharmonic on $\mathbb{B}$, then by Theorem 4.4.1 the function f^+ is also $\mathcal{H}$-subharmonic on $\mathbb{B}$.

Theorem 7.3.6 *Let f be $\mathcal{H}$-subharmonic on $\mathbb{B}$. Then the following are equivalent:*

(a) $\lim_{r\to 1} \int_{\mathbb{S}} f^+(rt)d\sigma(t) < \infty.$

(b) *$f(x) \leq P_h[\nu](x)$ for some signed Borel measure ν on $\mathbb{S}$.*

(c) *f has a least $\mathcal{H}$-harmonic majorant of the form $P_h[\nu_f]$ for some signed measure ν_f on $\mathbb{S}$.*

The measure ν_f of Theorem 7.3.6 is called the **boundary measure** of f.

Proof. Assume (a) holds. Since f^+ is $\mathcal{H}$-subharmonic, by Theorem 7.3.2 f^+ has a least $\mathcal{H}$-harmonic majorant H on $\mathbb{B}$. Since H is non-negative, by Theorem 7.1.1 $H(x) = P_h[\nu](x)$ for some non-negative measure ν on $\mathbb{S}$. Thus since

$$f(x) \leq f^+(x) \leq P_h[\nu](x),$$

we obtain (b).

Assume that (b) holds, that is, $f(x) \le P_h[\nu](x)$ for some signed measure ν on $\mathbb{S}$. Then for all r, $0 < r < 1$,

$$\int_{\mathbb{S}} f(rt)d\sigma(t) \le \int_{\mathbb{S}} P_h[\nu](rt)d\sigma(t) = P_h[\nu](0),$$

which is finite. Hence $\lim_{r\to 1^-} \int_{\mathbb{S}} f(rt)d\sigma(t) \le C < \infty$. Thus by Theorem 7.3.2 f has a least $\mathcal{H}$-harmonic majorant H_f on $\mathbb{B}$. Since H_f is the least $\mathcal{H}$-harmonic majorant of f,

$$H_f(x) \le P_h[\nu](x) \le P_h[\nu^+](x),$$

where ν^+ is the positive variation of ν. Therefore $H_f^+(x) \le P_h[\nu^+](x)$. Since $|H_f| = 2H_f^+ - H_f$,

$$\begin{aligned}\int_{\mathbb{S}} |H_f(rt)|d\sigma(t) &= 2\int_{\mathbb{S}} H_f^+(rt)d\sigma(t) - \int_{\mathbb{S}} H_f(rt)d\sigma(t)\\ &\le 2\nu^+(\mathbb{S}) - H_f(0).\end{aligned}$$

Therefore H_f satisfies the hypothesis of Theorem 7.1.1 and as a consequence, $H_f = P_h[\nu_f]$ for some signed Borel measure ν_f on $\mathbb{S}$. The implication (c) $\Rightarrow$ (a) is obvious. □

Remark 7.3.7 *If* $f(0) > -\infty$ *and* f *satisfies Theorem 7.3.6(a), then since* $|f(x)| = 2f^+(x) - f(x)$ *we have*

$$\sup_{0<r<1} \int_{\mathbb{S}} |f(rt)|d\sigma(t) \le 2\lim_{r\to 1} \int_{\mathbb{S}} f^+(rt)d\sigma(t) - f(0) < \infty.$$

Suppose f satisfies (a) of Theorem 7.3.6 and $F_f = P_h[\nu_f]$ is the least $\mathcal{H}$-harmonic majorant of f. Let g be continuous on $\mathbb{S}$, and set $G(x) = P_h[g](x)$. Since $P_h(r\zeta, t) = P_h(rt, \zeta)$ we have

$$\begin{aligned}\int_{\mathbb{S}} F_f(rt)g(t)d\sigma(t) &= \int_{\mathbb{S}} P_h[\nu_f](rt)g(t)d\sigma(t)\\ &= \int_{\mathbb{S}}\left[\int_{\mathbb{S}} P_h(rt,\zeta)g(t)d\nu_f(\zeta)\right] d\sigma(t)\\ &= \int_{\mathbb{S}}\left[\int_{\mathbb{S}} P_h(r\zeta,t)g(t)d\sigma(t)\right] d\nu_f(\zeta)\\ &= \int_{\mathbb{S}} G(r\zeta)d\nu_f(\zeta),\end{aligned}$$

where the right side tends to $\int_{\mathbb{S}} g(\zeta)d\nu_f(\zeta)$ as $r \to 1^-$. Hence $d\nu_f(t)$ is the weak limit of $F_f(rt)d\sigma(t)$. By (7.3.3) it now also follows that

$$\lim_{r\to 1} \int_{\mathbb{S}} F_f(rt)g(t)d\sigma(t) = \lim_{r\to 1^-} \int_{\mathbb{S}} f(rt)g(t)d\sigma(t) = \int_{\mathbb{S}} g(t)d\nu_f(t) \qquad (7.3.4)$$

for every $g \in C(\mathbb{S})$.

For the proof of Theorem 7.4.2 we require the following theorem.

Theorem 7.3.8 *Let $F = P_h[\nu]$ where ν is a finite signed measure on $\mathbb{S}$. If*

$$d\nu = \hat{f} d\sigma + d\nu_s$$

is the Lebesgue decomposition of ν where ν_s is singular with respect to σ and $\hat{f} \in L^1(\mathbb{S})$, then for every sequence $r_k \uparrow 1$, there exists a subsequence of $\{r_k\}$, denoted $\{r_k\}$, such that

$$\lim_{k\to\infty} F(r_k t) = \hat{f}(t) \quad \textit{for almost every } \ t \in \mathbb{S}.$$

Remark 7.3.9 *In Theorem 8.3.3 of the next chapter we prove a much stronger version of the above; namely, we prove that $\hat{f}$ is the non-tangential limit of F a.e. on $\mathbb{S}$. The much weaker version stated above is required for the proof of Theorem 7.4.2.*

For the proof of Theorem 7.3.8 we require the following lemma.

Lemma 7.3.10 *Let ν_s be singular with respect to σ and let $H(x) = P_h[\nu_s](x)$. Then $H_r \to 0$ in measure as $r \to 1$; that is, for every $\epsilon > 0$,*

$$\lim_{r\to 1} \sigma(\{t \in \mathbb{S} : |H(rt)| \geq \epsilon\}) = 0.$$

Proof. Since the total variation measure $|\nu_s|$ is also singular with respect to σ and $|H(x)| \leq P_h[|\nu_s|](x)$ we can without loss of generality assume that $\nu_s \geq 0$.

Since H_r converges weakly to ν_s we have

$$\lim_{r\to 1} \int_{\mathbb{S}} H(rt) g(t) d\sigma(t) = \int_{\mathbb{S}} g(t) d\nu_s(t)$$

for every $g \in C(\mathbb{S})$. Let C be a closed subset of $\mathbb{S}$. Since the characteristic function χ_C is upper semicontinuous, by Theorem 4.3.2 there exists a sequence $\{g_k\}$ of continuous functions such that $g_k \downarrow \chi_C$. Then

$$\limsup_{r\to 1} \int_C H_r d\sigma \leq \lim_{r\to 1} \int_{\mathbb{S}} H_r g_k d\sigma = \int_{\mathbb{S}} g_k(t) d\nu_s(t).$$

Since the above holds for all k we have

$$\limsup_{r\to 1} \int_C H_r d\sigma \leq \nu_s(C)$$

for every closed subset C of $\mathbb{S}$.

Since ν_s is singular, there exists a measurable set $A \subset \mathbb{S}$ such that $\sigma(A) = 0$ and $\nu_s(\mathbb{S} \setminus A) = 0$. Let U be any open subset of $\mathbb{S}$ containing A. Then for any $\epsilon > 0$,

$$\sigma(\{t \in \mathbb{S} \setminus U : H(rt) \geq \epsilon\}) \leq \frac{1}{\epsilon} \int_{\mathbb{S}\setminus U} H(rt) d\sigma(t).$$

Since $\mathbb{S} \setminus U$ is closed and contained in $\mathbb{S} \setminus A$,

$$\limsup_{r\to 1} \sigma(\{t \in \mathbb{S} \setminus U : H(rt) \geq \epsilon\}) = 0. \tag{7.3.5}$$

Since (7.3.5) is valid for all $\epsilon > 0$ and all open sets U containing A, $H_r \to 0$ in measure on $\mathbb{S} \setminus A$ and hence on $\mathbb{S}$. □

Proof of Theorem 7.3.8 Let $\{r_k\}$ be an increasing sequence in $(0, 1)$ with $r_k \to 1$. By Exercise 5.7.1, $P_h[\hat{f}](rt) \to \hat{f}(t)$ in L^1, and hence also in measure. Thus by [71, Proposition 18] there exists a subsequence of $\{r_k\}$, which without loss of generality is denoted by $\{r_k\}$, such that

$$\lim_{k\to\infty} P_h[\hat{f}](r_k t) = \hat{f}(t) \quad \text{and} \quad \lim_{k\to\infty} P_h[\nu_s](r_k t) = 0$$

for almost every $t \in \mathbb{S}$. Therefore

$$\lim_{k\to\infty} F(r_k t) = \hat{f}(t)$$

for almost every $t \in \mathbb{S}$. □

7.4 Hardy–Orlicz Spaces of $\mathcal{H}$-Subharmonic Functions

If f is $\mathcal{H}$-subharmonic and ϕ is a convex, non-decreasing function on $\mathbb{R}$ with $\phi(-\infty) = \lim_{t\to-\infty} \phi(t)$, then $\phi(f)$ is $\mathcal{H}$-subharmonic. A non-negative convex function ϕ is **strongly convex** if ϕ is non-decreasing, $\phi(t) \to \phi(-\infty)$ as $t \to -\infty$, and $\phi(t)/t \to \infty$ as $t \to \infty$.

Definition 7.4.1 *For a strongly convex function φ, set*

$$\mathcal{S}_\varphi = \left\{ f : f \text{ is } \mathcal{H}\text{-subharmonic and } \sup_{0<r<1} \int_{\mathbb{S}} \varphi(f(rt)) d\sigma(t) < \infty \right\}.$$

Likewise

$$\mathcal{H}_\varphi = \{ f : f \text{ is } \mathcal{H}\text{-harmonic and } |f| \in \mathcal{S}_\varphi \}.$$

The spaces $\mathcal{S}_\varphi$ and $\mathcal{H}_\varphi$ are called the **Hardy–Orlicz** spaces of $\mathcal{H}$-subharmonic and $\mathcal{H}$-harmonic functions corresponding to φ.

Since we assume that φ is non-negative, in the definition of $\mathcal{S}_\varphi$ we do not assume that the $\mathcal{H}$-subharmonic function is non-negative. If for $p > 1$ we take

$$\varphi_p(x) = \begin{cases} x^p & \text{if } x \geq 0, \\ 0 & \text{if } x < 0, \end{cases}$$

then φ_p is strongly convex and

$$\mathcal{S}_{\varphi_p} = \{f : f \text{ is } \mathcal{H}\text{-subharmonic with } f^+ \in \mathcal{S}^p\}.$$

On the other hand $\mathcal{H}_{\varphi_p} = \mathcal{H}^p$. We are now ready to prove the following analogue for $\mathcal{H}$-subharmonic functions of a theorem of L. Garding and L. Hörmander [26].

Theorem 7.4.2 *Let f be $\mathcal{H}$-subharmonic on $\mathbb{B}$ and let ϕ be a strongly convex function on $\mathbb{R}$. Assume that*

$$\int_{\mathbb{S}} \phi(f(rt))d\sigma(t) \le C < \infty, \quad 0 \le r < 1. \tag{7.4.1}$$

Then f has a boundary measure $d\nu_f = \hat{f}d\sigma + d\nu_s$ with $\hat{f} \in L^1$ and $\nu_s \le 0$. Furthermore, the boundary measure of $\phi(f)$ is absolutely continuous and equals $\phi(\hat{f})d\sigma$. In particular,

$$\lim_{r\to 1^-} \int_{\mathbb{S}} |\phi(f(rt)) - \phi(\hat{f}(t))|d\sigma(t) = 0. \tag{7.4.2}$$

Proof. Since $t/\phi(t) \to 0$ as $t \to \infty$, we have $f(rt) \le c\phi(f(rt))$ for some positive constant c. Since $\phi \ge 0$, we also have $f^+(rt) \le c\phi(f(rt))$. Hence by Theorem 7.3.6 f has a least $\mathcal{H}$-harmonic majorant of the form $P_h[\nu_f]$.

Let $O \subset \mathbb{S}$ be open and let $0 \le g \le 1$ be continuous and vanishing on $\mathbb{S} \setminus O$. Set

$$\alpha(s) = \sup(t/\phi(t)), \quad t \ge s > 0.$$

Then $f(rt) \le \alpha(s)\phi(f(rt))$ whenever $f(rt) \ge s$. Therefore

$$\int_{\mathbb{S}} f(rt)g(t)d\sigma(t) \le \alpha(s) \int_{\mathbb{S}} \phi(f(rt))g(t)d\sigma(t) + s \int_O g(t)d\sigma(t).$$

Hence by (7.3.4)

$$\int_{\mathbb{S}} g d\nu_f = \lim_{r\to 1^-} \int_{\mathbb{S}} f(rt)g(t)d\sigma(t) \le C\alpha(s) + s\sigma(O).$$

Since χ_O is lower semicontinuous, by Theorem 4.3.2 there exists an increasing sequence of continuous functions g_n with $g_n \to \chi_O$. Thus by the above

$$\nu_f(O) \le C\alpha(s) + s\sigma(O).$$

Set $s = \sigma(O)^{-1/2}$. Then $s \to \infty$ as $\sigma(O) \to 0$. Thus $\alpha(s) \to 0$ and the right side above goes to zero. Therefore

$$\lim_{\sigma(O)\to 0} \nu_f(O) \le 0$$

for every open set O.

Let $d\nu_f = \hat{f}d\sigma + d\nu_s$ be the Lebesgue decomposition of ν_f where $\hat{f}$ is in $L^1(\mathbb{S})$ and ν_s is singular with respect to σ. Since

$$\left|\int_{\mathbb{S}} g\hat{f}d\sigma\right| \leq \int_O |\hat{f}|d\sigma,$$

which goes to zero as $\sigma(O) \to 0$, we obtain

$$\lim_{\sigma(O)\to 0} \nu_f(O) = \lim_{\sigma(O)\to 0} \nu_s(O) \leq 0.$$

From this it now follows that $\nu_s \leq 0$.

Let ν be the boundary measure of $\phi(f)$. Then

$$\lim_{r\to 1^-} \int_{\mathbb{S}} \phi(f(rt))g(t)d\sigma(t) = \int_{\mathbb{S}} g(t)d\nu(t) \tag{7.4.3}$$

for every continuous function g. As a consequence of (7.3.3) and Theorem 7.3.8 there exists a sequence $r_j \to 1$ such that $f(r_j t) \to \hat{f}(t)$ a.e. Hence, taking $g \geq 0$ in (7.4.3) and applying Fatou's lemma we have

$$\int \phi(\hat{f})gd\sigma \leq \int gd\nu.$$

Hence $\phi(\hat{f}) \in L^1(\mathbb{S})$ and

$$\phi(\hat{f})d\sigma \leq d\nu.$$

Since $\nu_s \leq 0$,

$$f(x) \leq P_h[\hat{f}](x). \tag{7.4.4}$$

Hence by Jensen's inequality

$$\phi(f(x)) \leq P_h[\phi(\hat{f})](x). \tag{7.4.5}$$

Since $P_h[\nu]$ is the least $\mathcal{H}$-harmonic majorant of $\phi(f)$, we have by (7.4.5) that $d\nu \leq \phi(\hat{f})d\sigma$. Hence ν is absolutely continuous with $d\nu = \phi(\hat{f})d\sigma$. Finally, since

$$\begin{aligned}
&\lim_{r\to 1} \int_{\mathbb{S}} |\phi(f(rt)) - \phi(\hat{f}(t))|d\sigma(t) \\
&\quad \leq \lim_{r\to 1} \int_{\mathbb{S}} |P_h[\phi(\hat{f})](rt)d\sigma(t) - \phi(f(rt))|d\sigma(t) \\
&\quad + \lim_{r\to 1} \int_{\mathbb{S}} |P_h[\phi(\hat{f})](rt) - \phi(\hat{f}(t))|d\sigma(t),
\end{aligned}$$

the result (7.4.2) follows by (7.3.3) and Exercise 5.7.6. □

As a consequence of Theorem 7.4.2 we have the following theorem.

Theorem 7.4.3 *Suppose $f \geq 0$ is such that f^{p_o} is $\mathcal{H}$-subharmonic for some p_o, $0 < p_o \leq 1$. If $f \in \mathcal{S}^p$ for some $p > p_o$, then there exists a function $\hat{f} \in L^p(\mathbb{S})$ such that*

$$f^p(x) \leq \int_{\mathbb{S}} P_h(x,\zeta)\hat{f}^p(\zeta)d\sigma(\zeta) \tag{7.4.6}$$

and

$$\lim_{r\to 1}\int_{\mathbb{S}} |f^p(r\zeta) - \hat{f}^p(\zeta)|d\sigma(\zeta) = 0. \tag{7.4.7}$$

Proof. If $p_o = 1$, then we apply the previous theorem to f with $\phi(x) = x^p$, $x \geq 0$, which is strongly convex for $p > 1$. If $p_o < 1$, then we consider the $\mathcal{H}$-subharmonic function $g = f^{p_o}$ which is in $\mathcal{S}^r$ where $r = p/p_o$. By (7.4.4),

$$f^p(x) = g^r(x) \leq P_h[\hat{g}^r](x)$$

where $\hat{g}$ is the boundary function of g. Finally, since

$$\lim_{\rho\to 1}\int_{\mathbb{S}} |f^p(\rho\zeta) - \hat{g}^r(\zeta)|d\sigma(\zeta) = 0,$$

we have that $\hat{f}$ exists a.e. on $\mathbb{S}$ and $\hat{f}^p = \hat{g}^r$ a.e. on $\mathbb{S}$. □

Remark 7.4.4 *As a consequence of (7.4.7), if f^{p_o} is $\mathcal{H}$-subharmonic for some p_o, $0 < p_o \leq 1$, and $f \in \mathcal{S}^p$ for some $p > p_o$, then*

$$\|f\|_p^p = \lim_{r\to 1}\int_{\mathbb{S}} f^p(r\zeta)d\sigma(\zeta) = \|\hat{f}\|_p^p. \tag{7.4.8}$$

7.5 Exercises

7.5.1. **Harnack's theorem.** Let $\{f_n\}$ be a non-decreasing sequence of $\mathcal{H}$-harmonic functions on an open connected set $\Omega \subset \mathbb{B}$. Prove that either $f_n(x) \to \infty$ for all $x \in \Omega$ or that $\{f_n\}$ converges uniformly on compact subsets of Ω to an $\mathcal{H}$-harmonic function f.

7.5.2. (a) Fix $\zeta \in \mathbb{S}$. Prove that $P_h(\cdot,\zeta) \in \mathcal{H}^p$ for all p, $0 < p \leq 1$, but not for any $p > 1$.

(b) For $\alpha \geq 1$ show that $P_h^\alpha(\cdot,\zeta) \in \mathcal{H}^p_{\lambda_\alpha}$ for all p, $0 < p \leq \frac{1}{\alpha}$, where $\lambda_\alpha = 4(n-1)^2\alpha(\alpha-1)$.

7.5.3. (a) Let f be a non-negative $\mathcal{H}$-subharmonic function with least $\mathcal{H}$-harmonic majorant F_f. Prove that

$$f(x) \leq \left(\frac{1+|x|}{1-|x|}\right)^{n-1} F_f(0).$$

(b) Prove that

$$|f(x)|^p \le \left(\frac{1+|x|}{1-|x|}\right)^{n-1} \|f\|_p^p$$

for all $x \in \mathbb{B}$ and all $f \in \mathcal{H}^p$, $1 \le p < \infty$, and

$$|f(x)| \le \left(\frac{1+|x|}{1-|x|}\right)^{n-1} \|f\|_\infty, \quad p = \infty.$$

7.5.4. Prove that $(\mathcal{H}^\infty, d_\infty)$ is a complete metric space.

Definition 7.5.1 *A family of functions $\mathcal{F} \subset L^1(\mathbb{S})$ is said to be* **uniformly integrable** *if for every $\epsilon > 0$ there exists a $\delta > 0$ such that $\int_E |f| d\sigma < \epsilon$ whenever $f \in \mathcal{F}$ and $\sigma(E) < \delta$.*

7.5.5. (a) If $\mathcal{F} \subset \mathcal{S}_\varphi$ for some strongly convex function φ, prove that $\{f^+ : f \in \mathcal{F}\}$ is uniformly integrable on $\mathbb{S}$.
(b) Let f be $\mathcal{H}$-subharmonic on $\mathbb{B}$. If $\{f_r^+ : 0 < r < 1\}$ is uniformly integrable on $\mathbb{S}$, prove that $\nu_s \le 0$ where ν_s is the singular part of the boundary measure ν_f of f.

7.5.6. Let

$$H_*^1 = \{f \in \mathcal{H}^1 : \{|f_r|\}_{0<r<1} \quad \text{is uniformly integrable on } \mathbb{S}\}.$$

Prove that $f \in \mathcal{H}_*^1$ if and only if $f = P_h[\hat{f}]$ where $\hat{f} \in L^1(\mathbb{S})$.

Exercises on the Upper Half-Space $\mathbb{H}$

7.5.7. Suppose U is positive and $\mathcal{H}$-harmonic on $\mathbb{H}$. Prove that there exists a positive Borel measure μ on $\mathbb{R}^{n-1}$ and a non-negative constant c such that

$$U(x, y) = P_{\mathbb{H}}[\mu](x, y) + cy^{n-1}.$$

Hint: Consider $V(x) = U(\Phi(x))$ where Φ is the Möbius transformation given by 2.3.2. (See [10, Theorem 7.24] for the analogous result for Euclidean harmonic functions on $\mathbb{H}$.)

8

Boundary Behavior of Poisson Integrals

In Theorem 5.3.3 we proved that if f is a continuous function on $\mathbb{S}$, then $\lim_{x\to\zeta} P_h[f](x) = f(\zeta)$ for every $\zeta \in \mathbb{S}$. In this chapter we investigate the non-tangential boundary behavior of Poisson integrals of functions in $L^p(\mathbb{S})$, $1 \le p < \infty$. We begin with the Hardy–Littlewood maximal function in Section 8.1 for functions and measures on $\mathbb{S}$. The results of this section are standard and are included primarily for completeness. Section 8.2 contains the main results concerning radial and non-tangential maximal function of continuous functions on $\mathbb{B}$. In Section 8.3 we prove Fatou's theorem on the existence of non-tangential limits of Poisson integrals of measures, and in Section 8.4 we prove a local Fatou theorem for $\mathcal{H}$-harmonic functions. One of the results of Section 8.2 is the equivalence of the $L^p(\mathbb{S})$ norm of the radial and non-tangential maximal function for all p, $0 < p < \infty$. This result is then used in Section 8.5 to obtain an L^p inequality for the non-tangential maximal function valid for all p, $0 < p \le 1$. We close the chapter by giving an example of an $\mathcal{H}$-harmonic function that fails to have a finite radial limit at every $\zeta \in \mathbb{S}$.

8.1 Maximal Functions

For $\zeta \in \mathbb{S}$ and $\delta > 0$, let

$$S(\zeta,\delta) = \{t \in \mathbb{S} : |t - \zeta| < \delta\}. \tag{8.1.1}$$

Definition 8.1.1 *If μ is a signed Borel measure on $\mathbb{S}$, the* **maximal function** *of μ, denoted $M\mu$, is the function on $\mathbb{S}$ defined by*

$$M\mu(\zeta) = \sup_{\delta>0} \frac{|\mu|(S(\zeta,\delta))}{\sigma(S(\zeta,\delta))},$$

where $|\mu|$ *denotes the total variation of* μ. *Likewise, if* $f \in L^1(\mathbb{S})$, *the* **maximal function** *of* f, *denoted* Mf,[1] *is defined by*

$$Mf(\zeta) = \sup_{\delta>0} \frac{1}{\sigma(S(\zeta,\delta))} \int_{S(\zeta,\delta)} |f|\, d\sigma.$$

Theorem 8.1.2 *There exists a constant* B_n, *depending only on* n, *such that for any signed measure* μ *on* $\mathbb{S}$,

$$\sigma(\{\zeta \in \mathbb{S} : M\mu(\zeta) > t\}) \le B_n \frac{|\mu|(\mathbb{S})}{t}.$$

For the proof of Theorem 8.1.2 we require the following lemma.

Lemma 8.1.3 *Suppose* $E \subset \mathbb{S}$ *is the union of a finite collection* $\Phi = \{S(\zeta_i, \delta_i)\}$ *of balls in* $\mathbb{S}$. *Then there exists a finite disjoint collection* $\{S(\zeta_{i_k}, \delta_{i_k})\}_{k=1}^m$ *such that*

$$E \subset \bigcup_{k=1}^m S(\zeta_{i_k}, 3\delta_{i_k})$$

and

$$\sigma(E) \le B_n \sum_{k=1}^m \sigma(S(\zeta_{i_k}, \delta_{i_k})),$$

where B_n *is a constant depending only on* n.

Remark 8.1.4 *Although we prove Lemma 8.1.3 for a finite collection, with minor modifications, the result is also true if* Φ *is a countable collection. In this case, the conclusion states that there exists a finite or countable disjoint sub-collection of* Φ *such that the conclusions of the lemma hold.*

Proof. Without loss of generality we can assume that the balls $\{S(\zeta_i, \delta_i)\}$ have been ordered such that $\delta_{i+1} \le \delta_i$. Set $S_i = S(\zeta_i, \delta_i)$.

Suppose $\Gamma_k = \{S_{i_1}, \ldots, S_{i_k}\}$ have been chosen. If for each $i > i_k$, $S_i \cap S_{i_j} \ne \phi$ for some $j = 1, \ldots, k$, the process stops. If not, let $i_{k+1} > i_k$ be the smallest integer such that

$$S_{i_{k+1}} \bigcap S_{i_j} = \phi \quad \text{for all } j = 1, \ldots, k.$$

Set $\Gamma_{k+1} = \Gamma_k \cup \{S_{i_{k+1}}\}$. This process terminates since Φ is finite.

Let $S_i \in \Phi$ and let $\Gamma_m = \{S_{i_1}, \ldots, S_{i_m}\}$. If $i > i_m$, then $S_i \cap S_{i_j} \ne \phi$ for some i_j, $j \in \{1, \ldots, m\}$. Since $\delta_{i_j} \le \delta_i$, $S_i \subset 3S_{i_j}$.[2] If $i < i_m$, then there exists a k such

[1] The maximal function Mf is usually referred to as the Hardy–Littlewood maximal function. Also, in the definition of $M\mu$ and Mf one could have used δ^{n-1} instead of $\sigma(S(\zeta,\delta))$.

[2] For a ball $S = S(\zeta,\delta)$, $3S = S(\zeta, 3\delta)$.

that $i_k \leq i < i_{k+1}$. But then $S_i \cap S_{i_j} \neq \phi$ for some $j \in \{1, \ldots, k\}$. Since $\delta_{i_j} \geq \delta_i$ we have $S_i \subset 3S_{i_j}$. Thus

$$E \subset \bigcup_{j=1}^{m} 3S_{i_j}$$

and by the sub-additivity of the measure σ,

$$\sigma(E) \leq \sum_{j=1}^{m} \sigma(3S_{i_j}).$$

Finally, since $\sigma(S(\zeta, \delta)) \approx \delta^{n-1}$ (see Exercise 8.7.1), we have

$$\sigma(E) \leq B_n \sum_{j=1}^{m} \sigma(S(\zeta_{i_j}, \delta_{i_j})),$$

for some constant B_n. □

Proof of Theorem 8.1.2. We fix $t > 0$. We first note that for fixed $\delta > 0$ the function

$$\zeta \to \frac{|\mu|(S(\zeta, \delta))}{\sigma(S(\zeta, \delta))}$$

is lower semicontinuous on $\mathbb{S}$. Thus by Theorem 4.3.2 $M\mu(\zeta)$ is also lower semicontinuous on $\mathbb{S}$. Therefore

$$E(t) = \{\zeta \in \mathbb{S} : M\mu(\zeta) > t\}$$

is an open subset of $\mathbb{S}$. Let K be a compact subset of $E(t)$. For each $\zeta \in K$, there exists a $\delta_\zeta > 0$ such that

$$|\mu|(S(\zeta, \delta_\zeta)) > \sigma(S(\zeta, \delta_\zeta))\, t.$$

By compactness of K there exists a finite sub-collection $\{S(\zeta_i, \delta_i)\}$ which also covers K. Thus by Lemma 8.1.3 there exists a finite disjoint sub-collection $\{S(\zeta_{i_j}, \delta_{i_j})\}$ such that

$$\begin{aligned}
\sigma(K) &\leq B_n \sum_j \sigma(S(\zeta_{i_j}, \delta_{i_j})) \\
&\leq \frac{B_n}{t} \sum_j |\mu| \left(S(\zeta_{i_j}, \delta_{i_j})\right) \\
&\leq \frac{B_n}{t} |\mu| \left(\bigcup_j S(\zeta_{i_j}, \delta_{i_j})\right) \\
&\leq \frac{B_n}{t} |\mu|(S).
\end{aligned}$$

Taking the supremum over $K \subset E(t)$ proves the result. □

As a consequence of Theorem 8.1.2 we obtain the following corollary by defining the measure μ on $\mathbb{S}$ by $\mu(E) = \int_E f d\sigma$, $f \in L^1(\mathbb{S})$.

Corollary 8.1.5 *Let $f \in L^1(\mathbb{S})$. Then*

$$\sigma(\{\zeta \in \mathbb{S} : Mf(\zeta) > t\}) \leq \frac{C_n}{t} \int_{\mathbb{S}} |f| d\sigma.$$

Theorem 8.1.6 *For $1 < p \leq \infty$,*

$$\int_{\mathbb{S}} |Mf|^p d\sigma \leq A_p \int_{\mathbb{S}} |f|^p d\sigma$$

for all $f \in L^p(\mathbb{S})$, where $A_p = C_n \frac{p}{p-1}$ for $1 < p < \infty$ and $A_\infty = 1$.

For the proof of Theorem 8.1.6 we require the following lemma, the proof of which is left as an exercise (Exercise 8.7.2).

Lemma 8.1.7 *For $f \in L^p(\mathbb{S})$ we have*

$$\int_{\mathbb{S}} |f|^p d\sigma = p \int_0^\infty t^{p-1} \lambda(t) dt,$$

where

$$\lambda(t) = \sigma(\{\zeta \in \mathbb{S} : |f(\zeta)| > t\}).$$

Proof of Theorem 8.1.6. For the proof of this theorem we follow the method used by E. M. Stein in [75, Theorem 1, p. 5]. For $p = \infty$ we have $\|Mf\|_\infty \leq \|f\|_\infty$. Assume now that $1 < p < \infty$. For given $t > 0$ define f_1 by

$$f_1(x) = \begin{cases} f(x) & \text{if } |f(x)| \geq \frac{1}{2}t, \\ 0 & \text{otherwise.} \end{cases}$$

Then $|f(x)| \leq |f_1(x)| + \frac{1}{2}t$ and

$$\frac{1}{S(\zeta, \delta)} \int_{S(\zeta,\delta)} |f| d\sigma \leq Mf_1 + \frac{t}{2}.$$

Thus $Mf \leq Mf_1 + \frac{1}{2}t$, and by Corollary 8.1.5

$$\sigma(E_t) = \sigma(\{\zeta : Mf(\zeta) > t\}) \leq \frac{2C_n}{t} \|f_1\|_1,$$

or equivalently,

$$\sigma(E_t) = \sigma(\{\zeta : Mf(\zeta) > t\}) \leq \frac{2C_n}{t} \int_{|f| > t/2} |f| d\sigma. \tag{8.1.2}$$

Set $g = Mf$ and $\lambda(t) = \sigma(\{\zeta : g(\zeta) > t\})$. Then by Lemma 8.1.7

$$\int_{\mathbb{S}} (Mf)^p d\sigma = p \int_0^\infty t^{p-1} \lambda(t) dt.$$

But by (8.1.2) we have

$$\begin{aligned}\|Mf\|_p^p &= p\int_0^\infty t^{p-1}\lambda(t)dt \\ &\le p\int_0^\infty t^{p-1}\left(\frac{2C_n}{t}\int_{|f|>t/2}|f|d\sigma\right)dt,\end{aligned}$$

which by interchanging the order of integration

$$= 2pC_n\int_{\mathbb{S}}|f(\zeta)|\int_0^{2|f(\zeta)|}t^{p-2}dtd\sigma(\zeta).$$

But

$$\int_0^{2|f(\zeta)|}t^{p-2}dt = \frac{1}{(p-1)}|2f(x)|^{p-1}.$$

Therefore

$$\|Mf\|_p^p \le 2C_n\frac{p}{(p-1)}\int_{\mathbb{S}}|f|^p d\sigma,$$

from which the result now follows. □

Another application of Theorem 8.1.6 is the following differentiation theorem.

Theorem 8.1.8 *If $f \in L^1(\mathbb{S})$, then*

$$\lim_{\delta\to 0^+}\frac{1}{\sigma(S(\zeta,\delta))}\int_{S(\zeta,\delta)}|f - f(\zeta)|d\sigma = 0 \tag{8.1.3}$$

for almost every $\zeta \in \mathbb{S}$. Hence

$$f(\zeta) = \lim_{\delta\to 0^+}\frac{1}{\sigma(S(\zeta,\delta))}\int_{S(\zeta,\delta)}fd\sigma \qquad \text{a.e. on } \mathbb{S}. \tag{8.1.4}$$

Remark 8.1.9 *The points ζ for which (8.1.3) holds are called the* **Lebesgue points** *of f. If $E \subset \mathbb{S}$ and $f = \chi_E$, the characteristic function of E, then every point of E that is a Lebesgue point of f is called a* **point of density** *of E.*

Proof. If f is continuous, then (8.1.3) holds for every $\zeta \in \mathbb{S}$. For arbitrary $f \in L^1(\mathbb{S})$ and $\epsilon > 0$, choose $g \in C(\mathbb{S})$ such that $\|f - g\|_1 < \epsilon$.

Let

$$T_f(\zeta) = \limsup_{\delta\to 0^+}\frac{1}{\sigma(S(\zeta,\delta))}\int_{S(\zeta,\delta)}|f - f(\zeta)|d\sigma.$$

Then

(a) $T_f(\zeta) \le T_g(\zeta) + T_{f-g}(\zeta)$, and
(b) $T_{f-g}(\zeta) \le |f(\zeta) - g(\zeta)| + M[f-g](\zeta)$,

where $M[f-g]$ is the maximal function of $f-g$. Since g is continuous, $T_g(\zeta)=0$ for all $\zeta\in\mathbb{S}$. Thus

$$T_f(\zeta)\le|f(\zeta)-g(\zeta)|+M[f-g](\zeta).$$

Thus for $t>0$,

$$\begin{aligned}\sigma(\{\zeta:T_f(\zeta)>t\})\le\sigma(\{\zeta:|f(\zeta)-g(\zeta)|>\tfrac{t}{2}\})\\+\sigma(\{\zeta:M[f-g](\zeta)>\tfrac{t}{2}\}).\end{aligned}$$

If we let $E_t=\{\zeta:|f(\zeta)-g(\zeta)|>\frac{t}{2}\}$ then

$$\|f-g\|_1\ge\int_{E_t}|f-g|d\sigma\ge\frac{t}{2}\sigma(E_t).$$

Therefore,

$$\sigma(\{\zeta:|f(\zeta)-g(\zeta)|>\tfrac{t}{2}\})\le\frac{2}{t}\|f-g\|_1\le\frac{2\epsilon}{t}.$$

Also, by Corollary 8.1.5

$$\sigma(\{\zeta:M[f-g](\zeta)>\tfrac{t}{2}\})\le\frac{2C_n}{t}\|f-g\|_1\le\frac{2C_n\,\epsilon}{t}.$$

Therefore,

$$\sigma(\{\zeta:T_f(\zeta)>t\})\le\frac{C}{t}\epsilon.$$

Since $\epsilon>0$ was arbitrary we have

$$\sigma(\{\zeta:T_f(\zeta)>t\})=0$$

for every $t>0$. Thus $T_f(\zeta)=0$ a.e. on $\mathbb{S}$ which proves (8.1.3) as well as (8.1.4). □

Definition 8.1.10 *If ν is a finite positive measure on $\mathbb{S}$, define the* **upper derivate** $\overline{D}\nu$ *of ν on $\mathbb{S}$ by*

$$\overline{D}\nu(\zeta)=\limsup_{\delta\to0^+}\frac{\nu(S(\zeta,\delta))}{\sigma(S(\zeta,\delta))}. \tag{8.1.5}$$

If the limit in (8.1.5) exists, we write

$$D\nu(\zeta)=\lim_{\delta\to0^+}\frac{\nu(S(\zeta,\delta))}{\sigma(S(\zeta,\delta))}.$$

One of the advantages in considering $\overline{D}\nu$ is that unlike $M\nu$ it depends only on the behavior of $\nu(S(\zeta,\delta))$ for small values of δ.

Lemma 8.1.11 *If ν is a finite positive singular measure on $\mathbb{S}$, then*

$$D\nu = 0 \quad \sigma\text{- a.e. on } \mathbb{S}.$$

Proof. Suppose the conclusion is false. Then since $\overline{D}\mu$ is lower semicontinuous on $\mathbb{S}$, there exists a positive constant a, a compact set $K \subset \mathbb{S}$ with $\sigma(K) > 0$ and $\nu(K) = 0$, such that $\overline{D}\nu(\zeta) > a$ for all $\zeta \in K$.

Let $\epsilon > 0$ be arbitrary. Choose an open set $O \subset \mathbb{S}$ such that $K \subset O$ and $\nu(O) < \epsilon$. Since $\overline{D}\nu(\zeta) > a$ for every $\zeta \in K$, we can cover K by finitely many $S(\zeta_i, \delta_i) \subset O$ and $\nu(S(\zeta_i, \delta_i)) > a\sigma(S(\zeta_i, \delta_i))$. By Lemma 8.1.3 there exists a finite disjoint sub-collection $S(\zeta_{i_k}, \delta_{i_k})$ such that

$$\begin{aligned}\sigma(K) &\le B_n \sum_k \sigma(S(\zeta_{i_k}, \delta_{i_k}))\\ &\le \frac{B_n}{a} \sum_k \nu(S(\zeta_{i_k}, \delta_{i_k}))\\ &\le \frac{B_n}{a}\nu(O) \le \frac{B_n}{a}\epsilon.\end{aligned}$$

Since $\epsilon > 0$ was arbitrary, we have $\sigma(K) = 0$, which is a contradiction. □

8.2 Non-tangential and Radial Maximal Function

As in the Euclidean case, for $\zeta \in \mathbb{S}$ and $\alpha > 1$, we denote by $\Gamma_\alpha(\zeta)$ the **non-tangential approach region** at ζ defined by

$$\Gamma_\alpha(\zeta) = \{y \in \mathbb{B} : |y - \zeta| < \alpha(1 - |y|)\}. \tag{8.2.1}$$

Definition 8.2.1 *Let f be a continuous function on $\mathbb{B}$. For $\alpha > 1$, the* **non-tangential maximal function** *of f, denoted $M_\alpha f$, is defined on $\mathbb{S}$ by*

$$M_\alpha f(\zeta) = \sup\{|f(x)| : x \in \Gamma_\alpha(\zeta)\}.$$

Also, the **radial maximal function** *of f, denoted $M_{rad} f$, is the function on $\mathbb{S}$ defined by*

$$M_{rad} f(\zeta) = \sup_{0 \le r < 1} |f(r\zeta)|.$$

One clearly has that for any $f \in C(\mathbb{B})$, $M_{rad} f \le M_\alpha f$. In the reverse direction we have the following theorem.

Theorem 8.2.2 *For $0 < p < \infty$, $\alpha > 1$, there exists a constant $C_{p,\alpha}$ such that*

$$\int_{\mathbb{S}} (M_\alpha f)^p d\sigma \le C_{p,\alpha} \int_{\mathbb{S}} (M_{rad} f)^p d\sigma$$

for all non-negative, continuous, quasi-nearly $\mathcal{H}$-subharmonic functions.

The proof of Theorem 8.2.2 is similar to the proof given for harmonic functions in [27, Theorem 3.6] as well as Lemma 8 of [90].

Proof. Throughout the proof we let $E(x) = E(x, \frac{1}{2}) = \{y : |\varphi_x(y)| < \frac{1}{2}\}$. By Theorem 2.2.2, $E(x)$ is also a Euclidean ball $B(c_x, r_x)$ with center c_x and radius r_x given by

$$c_x = \frac{3x}{(4-|x|^2)} \quad \text{and} \quad r_x = \frac{2(1-|x|^2)}{(4-|x|^2)}.$$

Hence if $y \in E(x)$ we have

$$|y| \le |c_x| + r_x \le \frac{2|x|+1}{2+|x|} < 1$$

for all x, $|x| < 1$. Also, if we write $x = r\eta$ and $y = \rho t$, then

$$|t-\eta| \le (1-|y|) + (1-|x|) + |x-y|.$$

But

$$|x-y| \le 2r_x < \frac{4}{3}(1-|x|^2) \le \frac{8}{3}(1-|x|).$$

Also, by Exercise 2.4.1(b),

$$(1-|y|) \le 3(1-|x|^2) \le 6(1-|x|).$$

Therefore $|t-\eta| < c(1-|x|)$ for a fixed positive constant $c > 0$. Thus

$$E(x) \subset \left[0, \frac{2|x|+1}{2+|x|}\right] \times S(\eta, c(1-|x|)). \tag{8.2.2}$$

Let $x \in \Gamma_\alpha(\zeta), \zeta \in \mathbb{S}$, and let $0 < q < p$. Since f is quasi-nearly $\mathcal{H}$-subharmonic on $\mathbb{B}$,

$$f^q(x) \le 2^n C \int_{E(x)} f^q(y)\,d\tau(y),$$

which by (8.2.2)

$$\le C_n \int_0^{(2|x|+1)/(2+|x|)} (1-r)^{-n} \int_{S(\eta,c(1-|x|))} f^q(rt)\,d\sigma(t)\,dr.$$

Hence by Fubini's theorem,

$$f^q(x) \le C_n \int_{S(\eta,c(1-|x|))} (M_{rad} f)^q(t)\,d\sigma(t) \int_0^{(2|x|+1)/(2+|x|)} (1-r)^{-n}\,dr.$$

But

$$\int_0^{(2|x|+1)/(2+|x|)} (1-r)^{-n}\,dr \le \frac{c_n}{(1-|x|)^{n-1}}$$

and $S(\eta, c(1-|x|)) \subset S(\zeta, c_\alpha(1-|x|))$ for some fixed constant $c_\alpha > 0$. Therefore

$$f^q(x) \le \frac{C_{n,\alpha}}{(1-|x|)^{n-1}} \int\limits_{S(\zeta, c_\alpha(1-|x|))} (M_{rad}f)^q(t)d\sigma(t) \le Mg(\zeta),$$

where $g = (M_{rad}f)^q$. Therefore

$$(M_\alpha f)^p(\zeta) \le C_{n,\alpha}(Mg)^{p/q}(\zeta).$$

Since $p/q > 1$, by Theorem 8.1.6

$$\int_{\mathbb{S}} (M_\alpha f)^p d\sigma \le C_{p,\alpha,n} \int_{\mathbb{S}} g^{p/q} d\sigma = C_{p,\alpha,n} \int_{\mathbb{S}} (M_{rad}f)^p d\sigma,$$

provided $M_{rad}f \in L^p(\mathbb{S})$. □

Theorem 8.2.3 *For every $\alpha > 1$, there exists a constant A_α such that*

$$M_\alpha P_h[\mu] \le A_\alpha M\mu$$

for every signed Borel measure μ on $\mathbb{S}$.

The proof of the Theorem 8.2.3 follows standard techniques. A proof of the analogous result for the $\mathcal{M}$-harmonic Poisson kernel $\widetilde{\mathcal{P}}$ on the unit ball in $\mathbb{C}^n$ can be found in [72], whereas for the Euclidean Poisson kernel a proof of the result may be found in [10].

Proof. Since $|P_h[\mu]| \le P_h[|\mu|]$ we can assume without loss of generality that μ is a finite Borel measure on $\mathbb{S}$. Fix a $\zeta \in \mathbb{S}$ for which $M\mu(\zeta) < \infty$, and fix $x \in \Gamma_\alpha(\zeta)$. Set $\beta = \alpha(1-|x|)$. Let N be the smallest integer such that $2^n\beta > 2$. Set

$$V_0 = \{\tau \in \mathbb{S} : |\tau - \zeta| < \beta\},$$

and for $k = 1, \ldots, N$ set

$$V_k = \{\tau \in \mathbb{S} : 2^{k-1}\beta \le |\tau - \zeta| < 2^k\beta\}.$$

Then

$$\begin{aligned} P_h[\mu](x) &= \int_{\mathbb{S}} P_h(x,\tau)d\mu(\tau) \\ &= \int_{V_0} P_h(x,\tau)d\mu(\tau) + \sum_{k=1}^{N} \int_{V_k} P_h(x,\tau)d\mu(\tau). \end{aligned}$$

For $\tau \in V_0$ we have

$$P_h(x,\tau) = \frac{(1-|x|^2)^{n-1}}{|x-\tau|^{2(n-1)}} \le \frac{2^{n-1}}{(1-|x|)^{n-1}}.$$

Therefore,

$$\int_{V_0} P_h(x,\tau)d\mu(\tau) \le \frac{(2\alpha)^{n-1}}{\beta^{n-1}} \int_{S(\zeta,\beta)} d\mu \le (2\alpha)^{n-1} M\mu(\zeta).$$

For $x \in \Gamma_\alpha(\zeta)$ we have

$$\begin{aligned} |\tau-\zeta| &\le |x-\zeta| + |x-\tau| \\ &< \alpha(1-|x|) + |x-\tau| < (\alpha+1)|x-\tau|. \end{aligned}$$

Therefore $|x-\tau| > c_\alpha|\tau-\zeta|$, which for $\tau \in V_k$ gives

$$|x-\tau| > c_\alpha 2^{k-1}\beta$$

and thus

$$\begin{aligned} P_h(x,\tau) &\le \frac{2^{n-1}(1-|x|)^{n-1}}{c_\alpha^{2(n-1)}(2^{k-1})^{2(n-1)}\beta^{2(n-1)}} \\ &\le \frac{C_{\alpha,n}}{(2^{n-1})^k(2^k\beta)^{n-1}}. \end{aligned}$$

Therefore

$$\int_{V_k} P_h(x,\tau)d\mu(\tau) \le \frac{C_{\alpha,n}}{(2^{n-1})^k} \frac{1}{(2^k\beta)^{n-1}} \int\limits_{S(\zeta,2^k\beta)} d\mu \le \frac{C_{\alpha,n}}{(2^{n-1})^k} M\mu(\zeta).$$

Summing the above over k gives that for $x \in \Gamma_\alpha(\zeta)$,

$$P_h[\mu](x) = \int_{\mathbb{S}} P_h(x,\tau)d\mu(\tau) \le C_{n,\alpha}M\mu(\zeta).$$

Thus $M_\alpha P_h[\mu](\zeta) \le C_{n,\alpha}M\mu(\zeta)$. □

As a consequence of Theorem 8.2.3 we always have that

$$M_{rad}\, P_h[\mu] \le A\, M\mu \tag{8.2.3}$$

for any signed Borel measure μ on $\mathbb{S}$. We now prove that the reverse inequality also holds for positive Borel measures on $\mathbb{S}$.

Theorem 8.2.4 *If μ is a positive Borel measure on $\mathbb{S}$, then*

$$M\mu \le C_n M_{rad}\, P_h[\mu]$$

for a positive constant C_n depending only on n.

Proof. Fix $\zeta \in \mathbb{S}$ and for $0 < \delta < 2$ set $(1 - r) = \frac{1}{2}\delta$. Then

$$P_h[\mu](r\zeta) = \int_{\mathbb{S}} \frac{(1 - r^2)^{n-1}}{|r\zeta - \eta|^{2(n-1)}} d\sigma(\eta)$$
$$\geq \int_{S(\zeta,\delta)} \frac{(1 - r)^{n-1}}{|r\zeta - \eta|^{2(n-1)}} d\sigma(\eta).$$

But for $\eta \in S(\zeta,\delta)$, $|r\zeta - \eta| \leq (1 - r) + |\zeta - \eta| < \frac{3}{2}\delta$. Therefore

$$P_h[\mu](r\zeta) \geq \frac{2^{n-1}}{3^{2(n-1)}} \frac{1}{\delta^{n-1}} \mu(S(\zeta,\delta)).$$

Since $\sigma(S(\zeta,\delta)) \approx \delta^{n-1}$ (see Exercise 8.7.1)

$$P_h[\mu](r\zeta) \geq C_n \frac{\mu(S(\zeta,\delta))}{\sigma(S(\zeta,\delta))},$$

from which the result now follows. □

Thus if $\mu \geq 0$ and if one of $M\mu$, $M_{rad} P_h[\mu]$, $M_\alpha P_h[\mu]$ is finite at some $\zeta \in \mathbb{S}$, then so are the others.

Combining Theorems 8.1.6 and 8.2.3 gives the following theorem.

Theorem 8.2.5 *If $1 < p < \infty$ and $f \in L^p(\mathbb{S})$, then*

$$\int_{\mathbb{S}} |M_\alpha P_h[f]|^p d\sigma \leq A_{\alpha,p} \int_{\mathbb{S}} |f|^p d\sigma,$$

where $A_{\alpha,p}$ is a constant independent of f.

As a corollary we have the following characterization of functions in $\mathcal{S}^p$, and thus also $\mathcal{H}^p$.

Corollary 8.2.6 *Suppose $f \geq 0$ is such that f^{p_o} is $\mathcal{H}$-subharmonic for some p_o, $0 < p_o \leq 1$. Then $f \in \mathcal{S}^p$ for some $p > p_o$ if and only if $M_\alpha f \in L^p(\mathbb{S})$. If this is the case, then*

$$\|f\|_p \leq \|M_\alpha f\|_p \leq A_{p,\alpha} \|f\|_p.$$

Proof. Since $|f(r\zeta)|^p \leq (M_\alpha f)^p(\zeta)$, we have $f \in \mathcal{S}^p$ whenever $M_\alpha f \in L^p$ with $\|f\|_p \leq \|M_\alpha f\|_p$. Suppose now that $f \in \mathcal{S}^p$ for some $p > p_o$. Set $g = f^{p_o}$. Then $g \in \mathcal{S}^r$ with $r = p/p_o$. By inequality (7.4.4), $g(x) \leq P_h[\hat{g}](x)$ where $\hat{g}$ is the boundary function of g. Thus

$$f^p(x) = g^r(x) \leq (P_h[\hat{g}])^r(x),$$

and by Theorem 8.2.5

$$\int_{\mathbb{S}} (M_\alpha f)^p d\sigma \leq \int_{\mathbb{S}} (M_\alpha P_h[\hat{g}])^r d\sigma \leq A_{\alpha,r} \int_{\mathbb{S}} \hat{g}^r d\sigma .$$

But $\hat{g}^r = \hat{f}^p$ a.e. and $\|\hat{f}\|_p = \|f\|_p$, which proves the result. □

Remark 8.2.7 *For $p > 1$ we have that $f \in \mathcal{H}^p$ if and only if $M_\alpha f \in L^p$. Using this fact it is possible to define $\mathcal{H}^p$ for $0 < p \leq 1$ by*

$$\mathcal{H}^p = \{f : f \text{ is } \mathcal{H}\text{-harmonic and } M_\alpha f \in L^p(\mathbb{S})\}. \tag{8.2.4}$$

For further details the reader is referred to the papers [42] by P. Jaming and [31] by S. Grellier and P. Jaming.

8.3 Fatou's Theorem

In this section we will prove Fatou's theorem[3] on the existence of non-tangential limits of Poisson integrals of measures. We begin with the following definition.

Definition 8.3.1 *A continuous function f on $\mathbb{B}$ has a* **non-tangential limit** *at $\zeta \in \mathbb{S}$ if there exists a real number L such that*

$$\lim_{\substack{x \to \zeta \\ x \in \Gamma_\alpha(\zeta)}} f(x) = L$$

for every $\alpha > 1$.

In Theorem 8.3.3 we prove that if f is the Poisson integral of a measure on $\mathbb{S}$, then f has a non-tangential limit at a.e. $\zeta \in \mathbb{S}$. We begin with the following theorem.

Theorem 8.3.2 *If ν is a Borel measure on $\mathbb{S}$ satisfying $D\nu(\zeta) = 0$, $\zeta \in \mathbb{S}$, then for every $\alpha > 1$,*

$$\lim_{\substack{x \to \zeta \\ x \in \Gamma_\alpha(\zeta)}} P_h[\nu](x) = 0.$$

Proof. Let $\epsilon > 0$ be arbitrary. Since $D\nu(\zeta) = 0$, there exists a $\delta_o > 0$ such that

$$\nu(S(\zeta, \delta)) < \epsilon\, \sigma(S(\zeta, \delta)) \tag{8.3.1}$$

[3] In the setting of rank one symmetric spaces, Fatou's theorem was proved originally by A. W. Knapp in [45].

for all δ, $0 < \delta < \delta_o$. Let $S_o = S(\zeta, \delta_o)$. Set

$$\nu_o = \nu_{|S_o} \quad \text{and } \nu_1 = \nu - \nu_o.$$

If $t \in \mathbb{S} \setminus S_o$ then $|t - \zeta| \geq \delta_o$. Also, as in the proof of Theorem 8.2.3, $|t - \zeta| \leq (\alpha + 1)|x - t|$. Therefore for $x \in \Gamma_\alpha(\zeta)$,

$$P_h(x,t) \leq \frac{(\alpha+1)^{2(n-1)}(1-|x|^2)^{n-1}}{|t-\zeta|^{2(n-1)}} \leq \left(\frac{\alpha+1}{\delta_o}\right)^{2(n-1)} (1-|x|^2)^{n-1}.$$

Thus

$$\lim_{\substack{x \to \zeta \\ x \in \Gamma_\alpha(\zeta)}} P_h[\nu_1](x) = 0.$$

Also, by (8.3.1), $M[\nu_o](\zeta) < \epsilon$. Therefore by Theorem 8.2.3,

$$\limsup_{\substack{x \to \zeta \\ x \in \Gamma_\alpha(\zeta)}} M_\alpha P_h[\nu_o](x) \leq A_\alpha \, \epsilon,$$

from which the result follows. □

As a consequence of Theorem 8.1.8 we have the following result concerning non-tangential boundary limits of Poisson integrals.

Theorem 8.3.3 (Fatou's Theorem)
(a) *If $f \in L^1(\mathbb{S})$, then for every $\alpha > 1$,*

$$\lim_{\substack{x \to \zeta \\ x \in \Gamma_\alpha(\zeta)}} P_h[f](x) = f(\zeta) \quad \sigma\text{- a.e. on } \mathbb{S}.$$

(b) *If ν is a signed Borel measure on $\mathbb{S}$ which is singular with respect to σ, then for every $\alpha > 1$,*

$$\lim_{\substack{x \to \zeta \\ x \in \Gamma_\alpha(\zeta)}} P_h[\nu](x) = 0 \quad \sigma\text{- a.e. on } \mathbb{S}.$$

Proof. We first prove part (b) of the theorem. Suppose ν is a signed Borel measure on $\mathbb{S}$ which is singular with respect to σ. Since the total variation measure $|\nu|$ is also singular, by Lemma 8.1.11, $D|\nu|(\zeta) = 0 \, \sigma$ - a.e. on $\mathbb{S}$. Thus by Theorem 8.3.2,

$$\lim_{\substack{x \to \zeta \\ x \in \Gamma_\alpha(\zeta)}} P_h[|\nu|](x) = 0 \quad \sigma\text{- a.e. on } \mathbb{S}.$$

Since $|P_h[\nu]| \leq P_h[|\nu|]$ we obtain (b).

For the proof of (a), if $f \in L^1(\mathbb{S})$, by Theorem 8.3.2

$$\lim_{\delta \to 0^+} \frac{1}{\sigma(S(\zeta,\delta))} \int_{S(\zeta,\delta)} |f - f(\zeta)| d\sigma = 0$$

for almost every $\zeta \in \mathbb{S}$. Fix such a ζ and define μ on $\mathbb{S}$ by

$$\mu(E) = \int_E |f - f(\zeta)| d\sigma$$

for all Borel subsets E of $\mathbb{S}$. Then $D\mu(\zeta) = 0$ and

$$|P_h[f](x) - f(\zeta)| \leq P_h[\mu](x).$$

But then by Theorem 8.3.2 again,

$$\lim_{\substack{x \to \zeta \\ x \in \Gamma_\alpha(\zeta)}} |P_h[f](x) - f(\zeta)| \leq \lim_{\substack{x \to \zeta \\ x \in \Gamma_\alpha(\zeta)}} P_h[\mu](x) = 0,$$

which completes the proof. □

8.4 A Local Fatou Theorem for $\mathcal{H}$-Harmonic Functions

As a consequence of Theorems 7.1.1 and 8.3.3, if f is a bounded $\mathcal{H}$-harmonic function on $\mathbb{B}$, then f has a non-tangential limit at almost every $\zeta \in \mathbb{S}$. In this section we will prove a local Fatou theorem of I. Privalov [68] for (weakly) non-tangentially bounded $\mathcal{H}$-harmonic functions on $\mathbb{S}$. The result of Privalov was first extended to several variables for harmonic functions in half-spaces by A. P. Calderón [12]. See also [10, Chapter 7]. The result was subsequently proved by A. Koranyi and R. P. Putz [48] for harmonic functions on rank one symmetric spaces, which includes $\mathcal{H}$-harmonic functions in $\mathbb{B}$.

Our method of proof follows the techniques used, with minor modifications, by W. Rudin [72, Section 5.5] for $\mathcal{M}$-harmonic functions.

Definition 8.4.1 *For $E \subset \mathbb{S}$ and $\alpha > 1$, let*

$$\Omega(E, \alpha) = \bigcup_{\zeta \in E} \Gamma_\alpha(\zeta).$$

Our main result of this section is the following theorem.

Theorem 8.4.2 *If $E \subset \mathbb{S}$ is measurable with $\sigma(E) < 1$, $\alpha > 1$, and u is a bounded function on $\Omega(E, \alpha)$ such that $\Delta_h u = 0$, then*

$$\lim_{\substack{x \to \zeta \\ x \in \Gamma_\beta(\zeta)}} u(x) \quad \textit{exists}$$

for almost every $\zeta \in E$ and every $\beta > 1$.

For the proof of Theorem 8.4.2 we require several lemmas.

Lemma 8.4.3 *If $E \subset \mathbb{S}$, $\sigma(E) = m < 1$, $\alpha > 1$, then there exists a constant $c = c(\alpha, m) > 0$ such that*

$$P_h[\chi_{\mathbb{S}\setminus E}](x) \geq c$$

for all $x \in \mathbb{B} \setminus \Omega(E, \alpha)$.

Proof. Let $V = \mathbb{S} \setminus E$ and $f = \chi_V$. Let $x \in \mathbb{B} \setminus \Omega(E, \alpha)$. Write $x = r\eta$, $\eta \in \mathbb{S}$. If $r \leq \frac{1}{\alpha}$, then

$$P_h(r\eta, \zeta) = \frac{(1-r^2)^{n-1}}{|r\eta - \zeta|^{2(n-1)}} \geq \left(\frac{1-r}{1+r}\right)^{n-1}.$$

Therefore,

$$\int_{\mathbb{S}\setminus E} P_h(r\eta, \zeta)\,d\sigma(\zeta) \geq \left(\frac{1-r}{1+r}\right)^{n-1} \sigma(V) \geq \left(\frac{\alpha-1}{\alpha+1}\right)^{n-1} (1-m).$$

Assume $\frac{1}{\alpha} < r < 1$, and let φ_x be the automorphism of $\mathbb{B}$ satisfying Theorem 2.1.2. Then $(f \circ \varphi_x)(\zeta) = \chi_{\varphi_x(V)}(\zeta)$. Hence

$$P_h[f](x) = P_h[f \circ \varphi_x](0) = \int_{\mathbb{S}} (f \circ \varphi_x)\,d\sigma = \sigma(\varphi_x(V)).$$

Let

$$G_\alpha = \{t \in \mathbb{S} : 1 + \tfrac{1}{\alpha} < |t - \eta| \leq 2\}.$$

Suppose $t \in G_\alpha$. Then

$$|x - t| = |r\eta - t| \geq |\eta - t| - (1 - r),$$

which since $r > \frac{1}{\alpha}$ and $t \in G_\alpha$

$$> 1 + \tfrac{1}{\alpha} - (1 - \tfrac{1}{\alpha}) = \tfrac{2}{\alpha}.$$

Set $\zeta = \varphi_x(t)$. By Exercise 2.4.3,

$$|x - \zeta| = \frac{(1 - |x|^2)}{|x - t|}.$$

Therefore,

$$|x - \zeta| < \tfrac{\alpha}{2}(1 - |x|^2) < \alpha(1 - |x|).$$

Hence $x \in \Gamma_\alpha(\zeta)$. Since $x \notin \Omega(E, \alpha)$, $\zeta \in V$. Therefore $\varphi_x(G_\alpha) \subset V$. Hence $G_\alpha \subset \varphi_x(V)$ and $\sigma(\varphi_x(V)) > \sigma(G_\alpha) > 0$. Therefore

$$P_h[f](x) \geq \sigma(G_\alpha) \geq c_\alpha > 0 \quad \text{for} \quad |x| > \tfrac{1}{\alpha}.$$

Hence $P_h[f](x) \geq c = c(\alpha, m)$ for all $x \in \mathbb{S} \setminus \Omega(E, \alpha)$. □

Lemma 8.4.4 *If $\zeta \in E$ is a point of density of E, $\alpha > 1$, and $\beta > 1$, then there exists $r = r(E, \zeta, \alpha, \beta) < 1$ such that*

$$\{|x| > r\} \cap \Gamma_\beta(\zeta) \subset \Omega(E, \alpha).$$

Proof. Without loss of generality assume that $\sigma(E) < 1$. Define the measure ν on the Borel subsets of $\mathbb{S}$ by

$$\nu(B) = \int_B |\chi_E(t) - \chi_E(\zeta)| d\sigma(t).$$

Since ζ is a Lebesgue point of E, $D\nu(\zeta) = 0$. Hence by Theorem 8.3.2,

$$\lim_{\substack{x \to \zeta \\ x \in \Gamma_\beta(\zeta)}} P_h[\nu](x) = 0.$$

But

$$|P_h[\chi_E](x) - \chi_E(\zeta)| \le P_h[\nu](x).$$

Therefore

$$\lim_{\substack{x \to \zeta \\ x \in \Gamma_\beta(\zeta)}} P_h[\chi_E](x) = 1 \quad \text{and} \quad \lim_{\substack{x \to \zeta \\ x \in \Gamma_\beta(\zeta)}} P_h[\chi_{\mathbb{S} \setminus E}](x) = 0. \tag{8.4.1}$$

Let $\{r_j\}$ be an increasing sequence with $r_j \uparrow 1$. If the conclusion of the lemma is false, then for each j, there exists $x_j \in \{|x| > r_j\} \cap \Gamma_\beta(\zeta)$ with $x_j \in \mathbb{B} \setminus \Omega(E, \alpha)$. But then by Lemma 8.4.3, $P_h[\chi_{\mathbb{S} \setminus E}](x_j) \ge c > 0$, which is a contradiction. $\square$

Lemma 8.4.5 *Let $0 < r < 1$ be fixed. If $\mathcal{F}$ is a uniformly bounded family of functions which are $\mathcal{H}$-harmonic on B_r, then $\mathcal{F}$ is equicontinuous on each compact subset K of B_r.*

Proof. Let K be a compact subset of B_r. Choose $\delta > 0$ such that $\varphi_x(\overline{B}_\delta) \subset B_r$ for all $x \in K$. Let h be a non-negative C^∞ radial function with $\operatorname{supp} h \subset B_\delta$ and

$$\int_{\mathbb{B}} h d\tau = 1.$$

Let $x \in K$ and $f \in \mathcal{F}$. Without loss of generality we may assume that $|f| \le 1$ for all $f \in \mathcal{F}$. Since f is $\mathcal{H}$-harmonic on B_r,

$$f(x) = \int_{\mathbb{S}} f(\varphi_x(\rho\zeta)) d\sigma(\zeta)$$

for all ρ, $0 < \rho < \delta$. Therefore

$$
\begin{aligned}
f(x) &= n \int_0^1 f(x)h(\rho)\frac{\rho^{n-1}}{(1-\rho^2)^n}d\rho \\
&= n \int_0^1 \frac{\rho^{n-1}}{(1-\rho^2)^n} \int_{\mathbb{S}} h(\rho\zeta)f(\varphi_x(\rho\zeta))d\sigma(\zeta)d\rho \\
&= \int_{\mathbb{B}} h(y)f(\varphi_x(y))\, d\tau(y) = \int_{\mathbb{B}} f(y)h(\varphi_x(y))\, d\tau(y).
\end{aligned}
$$

Therefore, for $x, y \in K$,

$$|f(x) - f(y)| \le \int_{\mathbb{B}} |h(\varphi_x(w)) - h(\varphi_y(w))| d\tau(y),$$

which since h is radial

$$= \int_{\mathbb{B}} |(h \circ \varphi_w)(x) - (h \circ \varphi_w)(y)|\, d\tau(y).$$

Since $(w, x) \to (h \circ \varphi_w)(x)$ is continuous on $\overline{B_r} \times K$, given $\epsilon > 0$, there exists a $\delta' > 0$, $0 < \delta' < \delta$, such that $|(h \circ \varphi_w)(x) - (h \circ \varphi_w)(y)| < \epsilon$ for all $x, y \in K$ with $\rho_h(x, y) < \delta'$ and all $w \in B_r$. Therefore

$$|f(x) - f(y)| < \epsilon\tau(B_r)$$

for all $f \in \mathcal{F}$ and $y \in K$ with $\rho_h(x, y) < \delta'$. Hence the family $\mathcal{F}$ is equicontinuous on K. □

Proof of Theorem 8.4.2 Without loss of generality we can assume that $0 \le u \le 1$. Also, by replacing α by a slightly smaller α we can assume u is continuous on $\overline{\Omega}(E, \alpha) \cap \mathbb{B}$. Set $\Omega = \Omega(E, \alpha)$. Since u is continuous on $\overline{\Omega} \cap \mathbb{B}$, by Tietze's extension theorem [71, p. 179], u extends to a continuous function on $\mathbb{B}$, also denoted u, with $0 \le u \le 1$.

Let $r_j \uparrow 1$. By Theorem 5.4.1 there exists $g_j \in C(\overline{B}_{r_j})$ such that $g_j = u$ on S_{r_j} and $\Delta_h g_j = 0$ in B_{r_j} with $0 \le g_j \le 1$. By Lemma 8.4.5 the family $\{g_j\}_{j \ge j_o}$ is equicontinuous on $B_{r_{j_o}}$. Hence by the Ascoli–Arzela theorem [71, p. 169] there exists a subsequence, denoted $\{g_j\}$, that converges uniformly on compact subsets of $\mathbb{B}$ to an $\mathcal{H}$-harmonic function g on $\mathbb{B}$.

Let $v = P_h[\chi_{\mathbb{S}\setminus E}]$. Then by Lemma 8.4.3, $v \ge c$ on $\mathbb{B} \cap \partial\Omega$. Also, $|u - g_j| \le 1$ in B_{r_j} with $u - g_j = 0$ on S_{r_j}. Therefore

$$|u - g_j| \le \frac{1}{c} v \quad \text{on} \quad \partial(\Omega \cap B_{r_j}). \tag{8.4.2}$$

By the maximum principle (8.4.2) holds at every $x \in \Omega \cap B_{r_j}$. Letting $j \to \infty$ gives

$$|u - g| \le \frac{1}{c} v \quad \text{in} \quad \Omega. \tag{8.4.3}$$

Since g is bounded, by Theorem 7.1.1, $g = P_h[\hat{g}]$ for some bounded function $\hat{g}$ on $\mathbb{S}$. Hence by Theorem 8.3.3, the non-tangential limit of g exists at almost every $\zeta \in \mathbb{S}$. If ζ is a point of density of E, then by (8.4.3) and (8.4.1)

$$\lim_{\substack{x \to \zeta \\ x \in \Gamma_\beta(\zeta)}} g(x) = \lim_{\substack{x \to \zeta \\ x \in \Gamma_\beta(\zeta)}} u(x).$$

Therefore g and u have the same non-tangential limits a.e. on E. □

Definition 8.4.6 *A function $f : \mathbb{B} \to \mathbb{R}$ is said to be (weakly)* **non-tangentially bounded** *at $\zeta \in \mathbb{S}$ if there exists an $\alpha > 1$ such that*

$$\sup_{x \in \Gamma_\alpha(\zeta)} |f(x)| < \infty. \tag{8.4.4}$$

Likewise, f is non-tangentially bounded at $\zeta \in \mathbb{S}$ if (8.4.4) holds for all $\alpha > 1$.

Note: The term "weakly" is used to emphasize that (8.4.4) holds only for one $\alpha > 1$.

Theorem 8.4.7 *If u is $\mathcal{H}$-harmonic in $\mathbb{B}$, E is a measurable subset of $\mathbb{S}$, and u is weakly non-tangentially bounded at every $\zeta \in E$, then the non-tangential limit of u exists for almost every $\zeta \in E$.*

Proof. For $i = 1, 2, 3, \ldots$, let

$$E_i = \{\zeta \in \mathbb{S} : |u(x)| \le i \text{ for all } x \in \Gamma_{1+\frac{1}{i}}(\zeta)\}.$$

Since u is continuous in $\mathbb{B}$, each E_i is closed. By Theorem 8.4.2, u has a non-tangential limit at almost every point of each E_i. Since $E \subset \bigcup_i E_i$ the result follows. □

8.5 An L^p Inequality for $M_\alpha f$ for $0 < p \le 1$

In [88] (see also [87]) the author proved that if Ω is a bounded domain in $\mathbb{R}^n$ with $C^{1,1}$ boundary and if f is harmonic on Ω, then for $0 < p \le 1$, $\alpha > 1$, and $t_o \in \Omega$ fixed, there exists a constant $C_{\alpha,p}$, depending only on p and α, such that

$$\int_{\partial\Omega} (M_\alpha f)^p(\zeta)\, dS(\zeta) \le C_{\alpha,p} \left[|f(t_o)|^p + \int_\Omega \delta(y)^{p-1} |\nabla f(y)|^p dy \right],$$

where $\delta(y)$ is the distance from y to $\partial\Omega$. In the case of the unit ball this result has also been proved by O. Djordjević and M. Pavlović [17]. In the present section, as an application of Theorem 8.2.2, we prove that the above inequality also holds for $\mathcal{H}$-harmonic functions on $\mathbb{B}$.

Theorem 8.5.1 *Let f be an $\mathcal{H}$-harmonic function on $\mathbb{B}$. Then for $\alpha > 1$, $0 < p \le 1$,*

$$\int_{\mathbb{S}} (M_\alpha f)^p \, d\sigma \le C_{\alpha,p} \left[|f(0)|^p + \int_{\mathbb{B}} (1 - |x|^2)^{p-1} |\nabla f(x)|^p d\nu(x) \right],$$

where $C_{\alpha,p}$ is a constant independent of f.

Proof. By the fundamental theorem of calculus, for fixed r, $0 < r < 1$, and $\zeta \in \mathbb{S}$,

$$|f(r\zeta) - f(0)| = \left| \int_0^r \langle \nabla f(t\zeta), \zeta \rangle \, dt \right| \le \int_0^r \frac{|\nabla^h f(t\zeta)|}{(1-t)} dt.$$

Let $r_k = 1 - \frac{1}{2^k}$, $k = 0, 1, 2, \dots$, and let N be the first integer such that $r_N > r$. Then

$$\begin{aligned} |f(r\zeta)| &\le |f(0)| + \sum_{k=1}^{N} \int_{r_{k-1}}^{r_k} \frac{|\nabla^h f(t\zeta)|}{(1-t)} \, dt \\ &\le |f(0)| + (\log 2) \sum_{k=1}^{N} \sup_{t \in [r_{k-1}, r_k)} |\nabla^h f(t\zeta)|. \end{aligned}$$

Hence for $0 < p \le 1$,

$$|f(r\zeta)|^p \le |f(0)|^p + C \sum_{k=1}^{N} \sup_{t \in [r_{k-1}, r_k)} |\nabla^h f(t\zeta)|^p.$$

For $N = 1, 2, 3, \dots$ and $\zeta \in \mathbb{S}$, set

$$S_N(\zeta) = |f(0)|^p + C \sum_{k=1}^{N} \sup_{t \in [r_{k-1}, r_k)} |\nabla^h f(t\zeta)|^p$$

and

$$S(\zeta) = |f(0)|^p + C \sum_{k=1}^{\infty} \sup_{t \in [r_{k-1}, r_k)} |\nabla^h f(t\zeta)|^p.$$

Then $S_N(\zeta) \uparrow S(\zeta)$ as $N \to \infty$ and

$$(M_{rad} f)^p(\zeta) \le S(\zeta)$$

for all $\zeta \in \mathbb{S}$ for which $S(\zeta) < \infty$.

For $x \in \mathbb{B}$ let $B(x) = B(x, \frac{1}{4}(1 - |x|^2))$. Since $|\nabla^h f(x)|^p$ is quasi-nearly $\mathcal{H}$-subharmonic for all $p > 0$,

$$|\nabla^h f(t\zeta)|^p \le C_n \int_{B(t\zeta)} |\nabla^h f(x)|^p \, d\tau(x). \tag{8.5.1}$$

If $x \in B(t\zeta)$, $t \in (r_{j-1}, r_j)$, then

$$|x - r_{j-1}\zeta| \le |x - t\zeta| + (r - r_{j-1}) < \frac{1}{2^{j-1}}.$$

Therefore,

$$B(t\zeta) \subset B_j(\zeta) = B(r_{j-1}\zeta, 2^{-j+1}).$$

It is easy to show (see Exercise 2.4.2) that there exists $\beta > 0$ such that $B_j(\zeta) \subset \Gamma_\beta(\zeta)$. For $j = 2, 3, 4, \ldots$, set

$$A_j = \left\{ x \in \Gamma_\beta(\zeta) : \frac{1}{2^{j+2}} < (1 - |x|) < \frac{1}{2^{j-2}} \right\}.$$

Thus $B_k(\zeta) \subset A_k$ for all k. Hence by (8.5.1),

$$\sup_{t \in [r_{k-1}, r_k)} |\nabla^h f(t\zeta)|^p \le C_n \int_{A_k} |\nabla^h f(x)|^p d\tau(x).$$

For each $i = 1, 2, 3, 4$, the sequence of sets $\{A_{4k+i}\}_{k=1}^\infty$ is pairwise disjoint. Hence by Exercise 8.7.3,

$$\begin{aligned} S_N(\zeta) &\le |f(0)|^p + C_n \sum_{k=1}^N \int_{A_k} |\nabla^h f(x)|^p d\tau(x) \\ &\le |f(0)|^p + 4C_n \int_{\bigcup_{k=1}^\infty A_k} |\nabla^h f(x)|^p d\tau(x) \\ &\le |f(0)|^p + 4C_n \int_{\Gamma_\beta(\zeta)} |\nabla^h f(x)|^p d\tau(x). \end{aligned}$$

Letting $N \to \infty$ gives

$$S(\zeta) \le |f(0)|^p + C_n \int_{\Gamma_\beta(\zeta)} |\nabla^h f(x)|^p d\tau(x).$$

Hence

$$\int_{\mathbb{S}} S(\zeta) d\sigma(\zeta) \le |f(0)|^p + C \int_{\mathbb{S}} \int_{\Gamma_\beta(\zeta)} |\nabla^h f(x)|^p d\tau(x) d\sigma(\zeta),$$

which by Fubini's theorem

$$= \int_{\mathbb{B}} |\nabla^h f(x)|^p \left[\int_{\mathbb{S}} \chi_{\widetilde{\Gamma}_\beta(x)}(\zeta) d\sigma(\zeta) \right] d\tau(x).$$

In the above, for $\beta > 1$ and $x \in \mathbb{B}$,

$$\widetilde{\Gamma}_\beta(x) = \{\zeta \in \mathbb{S} : x \in \Gamma_\beta(\zeta)\}.$$

But by Exercise 8.7.4, if $x = |x|\eta$, $\eta \in \mathbb{S}$, then there exists a positive constant $c > 0$, depending only on β, such that

$$\widetilde{\Gamma}_\beta(x) \subset S(\eta, c(1 - |x|^2)).$$

Therefore

$$\begin{aligned} \int_{\mathbb{S}} S(\zeta) d\sigma(\zeta) &\le |f(0)|^p + C \int_{\mathbb{B}} (1 - |x|^2)^{n-1} |\nabla^h f(x)|^p d\tau(x) \\ &= |f(0)|^p + C \int_{\mathbb{B}} (1 - |x|^2)^{p-1} |\nabla f(x)|^p d\nu(x). \end{aligned}$$

Hence if $\int_{\mathbb{B}} (1 - |x|^2)^{p-1} |\nabla f(x)|^p d\nu(x) < \infty$, then $S(\zeta)$ is finite a.e. on $\mathbb{S}$, and thus

$$\int_{\mathbb{S}} (M_{rad} f)^p(\zeta) d\sigma(\zeta) \le |f(0)|^p + \int_{\mathbb{B}} (1 - |x|^2)^{p-1} |\nabla f(x)|^p d\nu(x).$$

The conclusion now follows by Theorem 8.2.2. □

Remark 8.5.2 *In Section 10.7 we will prove that Theorem 8.5.1 is still valid whenever* $1 < p \le 2$.

8.6 Example

In this final section we include an example of an $\mathcal{H}$-harmonic function which fails to have a finite radial limit at every $\zeta \in \mathbb{S}$. Our proof follows the methods used by S. Axler, P. Bourdon, and W. Ramey in proving the analogous result for Euclidean harmonic functions on $\mathbb{B}$ [10, p. 160].

Lemma 8.6.1 *For* $\zeta \in \mathbb{S}$, $m = 1, 2, \ldots$, *let* $f_m(\zeta) = e^{im\zeta_1}$. *Then* $P_h[f_m] \to 0$ *uniformly on compact subsets of* $\mathbb{B}$ *as* $m \to \infty$.

Proof. For $g \in C(\mathbb{S})$ and $t \in (-1, 1)$, set

$$G(t) = (1 - t^2)^{\frac{n-3}{2}} \int_{\mathbb{S}_{n-1}} g(t, \sqrt{1 - t^2}\, \zeta) d\sigma_{n-1}(\zeta).$$

By (5.5.11),

$$\int_{\mathbb{S}} g f_m d\sigma = c_n \int_{-1}^{1} G(t)e^{imt}dt.$$

Thus by the Riemann–Lebesgue lemma ([71, p. 169]), $\lim_{m\to\infty} \int f_m g d\sigma = 0$. By taking $g(\zeta) = P_h(x,\zeta)$ we have $\lim_{m\to\infty} P_h[f_m](x) = 0$ for all $x \in \mathbb{B}$.

Since $|f_m| \equiv 1$ on $\mathbb{S}$, we have $|P_h[f_m](x)| \le 1$ on $\mathbb{B}$. Thus by Lemma 8.4.5 the family $\{P_h[f_m]\}$ is equicontinuous on each compact subset K of $\mathbb{B}$. Hence by the Ascoli–Arzela theorem [71, p. 169] every subsequence of $\{P_h[f_m]\}$ contains a subsequence converging uniformly on compact subsets of $\mathbb{B}$. Since $\{P_h[f_m]\}$ converges pointwise to 0, we have $P_h[f_m] \to 0$ uniformly on compact subsets of $\mathbb{B}$. □

Remark 8.6.2 *The above sequence $\{P_h[f_m]\}$ is interesting in that each function $P_h[f_m]$ extends continuously to $\mathbb{S}$ with boundary values of modulus one everywhere on $\mathbb{S}$, yet converges uniformly to zero on compact subsets of $\mathbb{B}$.*

Theorem 8.6.3 *Let $\alpha : [0,1) \to [1,\infty)$ be an increasing function with $\alpha(r) \to \infty$ as $r \to 1$. Then there exists an $\mathcal{H}$-harmonic function on $\mathbb{B}$ such that*

(a) $|u(r\zeta)| < \alpha(r)$ *for all $r \in [0,1)$ and all $\zeta \in \mathbb{S}$;*
(b) *at every $\zeta \in \mathbb{S}$, $u(r\zeta)$ fails to have a finite limit as $r \to 1$.*

Proof. Choose an increasing sequence $\{s_m\} \subset [0,1)$ such that $\alpha(s_m) > m+1$. From the sequence $\{P_h[f_m]\}$ choose a subsequence $\{v_m\}$ such that $|v_m| < 2^{-m}$ on B_{s_m}. Suppose $r \in [s_m, s_{m+1})$. Since each v_m satisfies $|v_m| < 1$ on $\mathbb{B}$ we have

$$\begin{aligned}\sum_{k=1}^{\infty} |v_k(r\zeta)| &= \sum_{k=1}^{m} |v_k(r\zeta)| + \sum_{k=m+1}^{\infty} |v_k(r\zeta)| \\ &< m + \sum_{k=m+1}^{\infty} 2^{-k} \\ &< m+1 \\ &< \alpha(s_m) \\ &\le \alpha(r).\end{aligned}$$

Thus $\sum |v_k(r\zeta)| < \alpha(r)$ for all $r \in (0,1)$ and all $\zeta \in \mathbb{S}$. Furthermore the series $\sum |v_k|$ converges uniformly on compact subsets of $\mathbb{B}$.

From the sequence $\{v_m\}$ we inductively extract a subsequence $\{u_m\}$ in the following manner: set $u_1 = v_1$. Because u_1 is continuous on $\overline{\mathbb{B}}$ we may choose $r_1 \in [0, 1)$ such that

$$|u_1(r\zeta) - u_1(\zeta)| < \frac{1}{4}$$

for all $r \in [r_1, 1]$ and all $\zeta \in \mathbb{S}$. Suppose that we have chosen functions $u_1, u_2, \ldots, u_m$ from $\{v_m\}$ and radii $0 < r_1 < \cdots < r_m < 1$ such that

$$\sum_{k=1}^{m} |u_k(r\zeta) - u_k(s\zeta)| < \frac{1}{4} \quad \text{for all } r, s \in [r_m, 1],\ \zeta \in \mathbb{S}. \tag{8.6.1}$$

We now choose u_{m+1} such that $|u_{m+1}| < 2^{-(m+1)}$ on $\overline{B_{r_m}}$. Now choose $r_{m+1} \in (r_m, 1)$ so that (8.6.1) holds with $m + 1$ in place of m. The radius r_{m+1} can be chosen since u_{m+1} is continuous on $\overline{\mathbb{B}}$. We now define u as

$$u(x) = \sum_{m=1}^{\infty} u_m(x).$$

Since the series converges uniformly on compact subsets of $\mathbb{B}$, u is $\mathcal{H}$-harmonic on $\mathbb{B}$. Furthermore, since $|u| \le \sum |v_k|$ we have $|u(r\zeta)| < \alpha(r)$ for all $r \in [0, 1)$ and $\zeta \in \mathbb{S}$.

We now show that u fails to have a finite radial limit at each $\zeta \in \mathbb{S}$. For each $\zeta \in \mathbb{S}$ and each integer m we have

$$\begin{aligned}
|u(r_{m+1}\zeta) - u(r_m\zeta)| &\ge |u_{m+1}(r_{m+1}\zeta) - u_{m+1}(r_m\zeta)| \\
&\quad - \sum_{k\ne m+1} |u_k(r_{m+1}\zeta) - u_k(r_m\zeta)| \\
&\ge |u_{m+1}(\zeta)| - |u_{m+1}(r_{m+1}\zeta) - u_{m+1}(\zeta)| \\
&\quad - |u_{m+1}(r_m\zeta)| - \frac{1}{4} - 2\sum_{k=m+2}^{\infty} 2^{-k} \\
&\ge 1 - \frac{1}{4} - 2^{-(m+1)} - \frac{1}{4} - 2\sum_{k=m+2} 2^{-k} \\
&\ge \frac{1}{2} - 2\sum_{k=m+1} 2^{-k}.
\end{aligned}$$

In the above we have used the fact that $|u_{m+1}(\zeta)| = 1$ for all $\zeta \in \mathbb{S}$. Thus since $|u(r_{m+1}\zeta) - u(r_m\zeta)|$ is bounded away from zero, the sequence $\{u(r_m\zeta)\}$ fails to have a finite limit as $m \to \infty$. Hence $u(r\zeta)$ also fails to have a finite limit as $r \to 1$. □

8.7 Exercises

8.7.1. (a) Using the invariance of σ and (5.5.11) show that

$$\sigma(S(\zeta,\delta)) = c_n \int_{Q(\delta)} (1-s^2)^{\frac{n-3}{2}}\, ds$$

where $Q(\delta) = \{s : s < 1 \text{ and } 1 - s < \frac{1}{2}\delta^2\}$.

(b) Using the above show that there exist positive constants c_1 and c_2 such that

$$c_1\delta^{n-1} \leq \sigma(S(\zeta,\delta)) \leq c_2\delta^{n-1}$$

for all δ, $0 < \delta < \sqrt{2}$.

8.7.2. Let (X,μ) be a measure space. If f is a measurable function on X, define the distribution function λ by

$$\lambda(t) = \mu(\{x \in X : |f(x)| > t\}).$$

Prove that for $f \in L^p(X,\mu)$, $1 < p < \infty$,

$$\int_X |f|^p d\mu = p\int_0^\infty t^{p-1}\lambda(t)dt.$$

Hint: First prove the result for the case where s is a non-negative simple function given by

$$s(x) = \sum_{j=1}^{m} \alpha_j \chi_{E_j}(x),$$

where the $\{E_j\}$ are pairwise disjoint in X, and where without loss of generality $\alpha_i < \alpha_{i+1}$ for all i, $i = 1,\ldots,m-1$.

8.7.3. Let (X,μ) be a measure space and let $\{A_k\}$ be a sequence of measurable sets such that $\{A_{4k+i}\}_{k=0}^{\infty}$ are pairwise disjoint for each $i = 0,1,2,3$. Prove that for $f \in L^1(\mu)$,

$$\sum_{k=0}^{\infty} \int_{A_k} |f|\, d\mu \leq 4 \int_{\cup_{k=0}^{\infty} A_k} |f|\, d\mu.$$

8.7.4. For $\beta > 1$, $x \in \mathbb{B}$, let $\widetilde{\Gamma}_\beta(x) = \{\zeta \in S : x \in \Gamma_\beta(\zeta)\}$. Prove that there exists a constant $c > 0$, depending only on β such that if $x = |x|\eta$, $\eta \in \mathbb{S}$,

$$\widetilde{\Gamma}_\beta(x) \subset S(\eta, c(1-|x|^2)).$$

8.7.5. **Weighted radial limits of Poisson integrals.** Let ν be a non-negative singular measure on $\mathbb{S}$. Prove that

$$\lim_{r\to 1}(1-r)^{n-1}P_h[\nu](r\zeta) = 2^{n-1}\nu(\{\zeta\}) \quad \text{for every } \zeta \in \mathbb{S}.$$

The following exercise is a generalization of an inequality due to Hardy and Littlewood [35] for $n = 2$. A statement of the inequality, along with a proof using derivatives, may be found in [99, p. 101].

8.7.6. For $U \in L^1(\mathbb{S})$, $U \geq 0$, let

$$U^*(\zeta) = \sup_{0<\rho<1} \frac{1}{\rho^{n-1}} \int_{S(\zeta,\rho)} U(t)d\sigma(t).$$

If $U(x) = P_h[U](x)$, $x \in \mathbb{B}$, prove that there exists a constant C, independent of U, such that

$$U(r\eta) \leq CU^*(\zeta)\left[1 + \left(\frac{|\eta - \zeta|}{1-r}\right)^{n-1}\right].$$

Hint: Consider the two cases $|\eta - \zeta| > c(1-r)$ and $|\eta - \zeta| \leq c(1-r)$, where $c > 0$ is a constant.

8.7.7. For $\alpha > \frac{1}{2}$ and $\zeta \in \mathbb{S}$, let

$$\mathcal{A}_\alpha(\zeta) = \{x \in \mathbb{B} : |x - \zeta| < \alpha(1 - |x|^2)\}.$$

If U is a continuous function on $\mathbb{B}$, prove that U has a non-tangential limit L at $\zeta \in \mathbb{S}$ if and only if

$$\lim_{\substack{x \to \zeta \\ x \in \mathcal{A}_\alpha(\zeta)}} U(x) = L.$$

Exercises on the Upper Half-Space $\mathbb{H}$

8.7.8. For $\alpha > 0$ and $a \in \mathbb{R}^{n-1}$, set

$$\Gamma_\alpha(a) = \{(x, y) \in \mathbb{H} : |x - a| < \alpha y\}.$$

A function V on $\mathbb{H}$ has a non-tangential limit at a if and only if

$$\lim_{\substack{x \to a \\ x \in \Gamma_\alpha(a)}} V(x) \quad \text{exists.}$$

Prove that V has a non-tangential limit at $a \in \mathbb{R}^{n-1}$ if and only if $U(x) = V(\Phi(x))$ has a non-tangential limit at $\Phi(a)$.

9

The Riesz Decomposition Theorem for $\mathcal{H}$-Subharmonic Functions

One version of the classical Riesz decomposition theorem for subharmonic functions on the unit ball $\mathbb{B}$ is as follows: if $f \in C^2(\mathbb{B}) \cap C(\overline{\mathbb{B}})$ is subharmonic on $\mathbb{B}$, then

$$f(x) = H(x) - \int_{\mathbb{B}} G^e(x, y)\Delta f(y) d\nu(y),$$

where H is harmonic on $\mathbb{B}$ and G^e is the Euclidean Green's function on $\mathbb{B}$. For the invariant Laplacian Δ_h, by Theorem 4.1.1 we have

$$f(x) = (f \circ \varphi_x)(0) = \int_{\mathbb{S}} f(\varphi_x(rt)) d\sigma(t) - \int_{\mathbb{B}} g(|y|, r)\Delta_h(f \circ \varphi_x)(y) d\tau(y),$$

where

$$g(|y|, r) = \frac{1}{n} \int_{|y|}^{r} \frac{(1 - s^2)^{n-2}}{s^{n-1}} ds.$$

But by Theorem 5.3.5, if $f \in C(\overline{\mathbb{B}})$,

$$\lim_{r \to 1} \int_{\mathbb{S}} f(\varphi_x(rt)) d\sigma(t) = \int_{\mathbb{S}} P_h(x, t) f(t) d\sigma(t).$$

Also, $\lim_{r \to 1} g(|y|, r) = g_h(y)$, the Green's function on $\mathbb{B}$. Hence if $f \in C^2(\mathbb{B}) \cap C(\overline{\mathbb{B}})$, then

$$f(x) = \int_{\mathbb{S}} P_h(x, t) f(t) d\sigma(t) - \int_{\mathbb{B}} g_h(y)\Delta_h(f \circ \varphi_x)(y) d\tau(y),$$

which by the invariance of τ

$$= \int_{\mathbb{S}} P_h(x, t) f(t) d\sigma(t) - \int_{\mathbb{B}} G_h(x, y)\Delta_h f(y) d\tau(y).$$

In particular, the above is valid if f is $\mathcal{H}$-subharmonic on $\mathbb{B}$, in which case $\Delta_h f$ is non-negative on $\mathbb{B}$.

In Section 9.1 we extend the above to $\mathcal{H}$-subharmonic functions on $\mathbb{B}$ without the hypothesis that $f \in C^2(\mathbb{B}) \cap C(\overline{\mathbb{B}})$. Our proof follows the methods used by D. Ullrich in his dissertation [93], and published in [94], for the analogous results for $\mathcal{M}$-subharmonic functions on $\mathbb{B} \subset \mathbb{C}^n$. Since the results for $n = 2$ are well known and can be found elsewhere, we will assume throughout this chapter that $n \geq 3$. In Section 9.3 we extend a result of D. H. Armitage [8] concerning the integrability of superharmonic functions to $\mathcal{H}$-superharmonic functions on $\mathbb{B}$, and in Sections 9.4 and 9.5 we investigate boundary limits of Green potentials and non-tangential limits of $\mathcal{H}$-subharmonic functions.

9.1 The Riesz Decomposition Theorem

As a consequence of the above (see also Corollary 4.1.5) if f is C^2 in $\mathbb{B}$ with compact support, then

$$f(a) = -\int_{\mathbb{B}} G_h(a,x)\Delta_h f(x)d\tau(x). \tag{9.1.1}$$

In our next theorem we prove the following generalization of the above.

Theorem 9.1.1 *Suppose $f \leq 0$ is $\mathcal{H}$-subharmonic on $\mathbb{B}$ satisfying*

$$\lim_{r\to 1}\int_{\mathbb{S}} f(rt)\, d\sigma(t) = 0. \tag{9.1.2}$$

Then for all $x \in \mathbb{B}$,

$$f(x) = -\int_{\mathbb{B}} G_h(x,y)d\mu_f(y),$$

where μ_f is the Riesz measure of f.

For the proof of Theorem 9.1.1 we require the following lemma.

Lemma 9.1.2 *Suppose χ is a C^2 function with compact support which satisfies $\int_{\mathbb{B}} \chi\, d\tau = 0$. Let $v = -g_h * \chi$. Then*

$$v \in C^2_c(\mathbb{B}) \quad \textit{and} \quad \Delta_h v = \chi.$$

Proof. In the above, g_h is the Green's function on $\mathbb{B}$ given in (3.2.1) and $*$ is the invariant convolution on $\mathbb{B}$.

For fixed x, the function $y \to G_h(x,y)$ is $\mathcal{H}$-harmonic in $B(0,|x|)$. Thus for all ρ, $0 < \rho < |x|$,

$$\int_{\mathbb{S}} G_h(x,\rho t)d\sigma(t) = G_h(x,0) = g_h(x).$$

Choose r, $0 < r < 1$, such that χ has support in $\overline{B_r}$. Then for all x, $|x| > r$,

$$\begin{aligned}(g_h * \chi)(x) &= \int_{\mathbb{B}} G_h(x,y)\chi(y)d\tau(y)\\ &= n\int_0^r \rho^{n-1}(1-\rho^2)^{-n}\chi(\rho)\int_{\mathbb{S}} G_h(x,\rho t)d\sigma(t)d\rho\\ &= g_h(x)\int_{B_r}\chi(y)d\tau(y) = 0.\end{aligned}$$

Thus $\operatorname{supp} v \subset B_{|x|}$.

Let $\psi \in C_c^2(\mathbb{B})$ be arbitrary. Then by Green's formula (see (5.1.1)) and Fubini's theorem,

$$\begin{aligned}\int_{\mathbb{B}} \psi(y)\Delta_h v(y)d\tau(y) &= \int_{\mathbb{B}} v(y)\Delta_h\psi(y)d\tau(y)\\ &= -\int_{\mathbb{B}}\left[\int_{\mathbb{B}} G_h(x,y)\chi(x)d\tau(x)\right]\Delta_h\psi(y)d\tau(y)\\ &= -\int_{\mathbb{B}}\chi(x)\left[\int_{\mathbb{B}} G_h(x,y)\Delta_h\psi(y)d\tau(y)\right]d\tau(x),\end{aligned}$$

which by (9.1.1)

$$= \int_{\mathbb{B}} \chi(x)\psi(x)d\tau(x).$$

Therefore

$$\int_{\mathbb{B}} \psi(x)\Delta_h v(x)d\tau(x) = \int_{\mathbb{B}} \psi(x)\chi(x)d\tau(x).$$

Since v is C^2 and the above holds for all $\psi \in C_c^2(\mathbb{B})$ we have $\Delta_h v = \chi$. □

Proof of Theorem 9.1.1 Choose a sequence $\{r_j\}$, $0 < r_j < \frac{1}{2}$, which decreases to 0. Let

$$A_j^1 = \{x : r_{j+1} \le |x| \le r_j\} \quad \text{and} \quad A_j^2 = \{x : 1 - r_j \le |x| \le 1 - r_{j+1}\}.$$

For each $j = 1, 2, \dots$ and $k = 1, 2$, let $\chi_j^k \ge 0$ be a C^2 radial function with $\operatorname{supp}\chi_j^k \subset A_j^k$ and $\int \chi_j^k d\tau = 1$.

Since f is $\mathcal{H}$-subharmonic, by Theorem 4.5.4,

$$f(0) = \lim_{j\to\infty}(f * \chi_j^1)(0) = \lim_{j\to\infty}\int_{\mathbb{B}} f(x)\chi_j^1(x)d\tau(x).$$

Also, for each j, if $\rho_j = 1 - r_j$,

$$\int_{\mathbb{S}} f(\rho_j t)d\sigma(t) \le \int_{\mathbb{B}} f(x)\chi_j^2(x)d\tau(x) \le \int_{\mathbb{S}} f(\rho_{j+1}t)d\sigma(t).$$

Hence by our hypothesis on f,

$$\lim_{j\to\infty} \int_{\mathbb{B}} f(x)\chi_j^2(x)d\tau(x) = 0.$$

Hence if $\chi_j = \chi_j^1 - \chi_j^2$,

$$\lim_{j\to\infty} \int_{\mathbb{B}} f(x)\chi_j(x)d\tau(x) = f(0).$$

Furthermore, since the Green's function g_h is $\mathcal{H}$-superharmonic on $\mathbb{B}$, $g_h * \chi_j^1$ increases to g_h. Also, a similar argument as in Theorem 4.5.4 shows that $g_h * \chi_j^2$ decreases to zero. Thus $g_h * \chi_j$ increases to g_h.

For each j let $v_j = -(g_h * \chi_j)$. By Lemma 9.1.2, $v_j \in C_c^2(\mathbb{B})$ with $\Delta_h v_j = \chi_j$. Thus

$$\begin{aligned} f(0) &= \lim_{j\to\infty} \int_{\mathbb{B}} f(x)\chi_j(x)d\tau(x) \\ &= \lim_{j\to\infty} \int_{\mathbb{B}} f(x)\Delta_h v_j(x)d\tau(x), \end{aligned}$$

which by Theorem 4.6.3

$$= \lim_{j\to\infty} \int_{\mathbb{B}} v_j(x)d\mu_f(x),$$

where μ_f is the Riesz measure of f. Therefore,

$$f(0) = -\lim_{j\to\infty} \int_{\mathbb{B}} (g_h * \chi_j)(x)d\mu_f(x),$$

which by the monotone convergence theorem

$$= -\int_{\mathbb{B}} g_h(x)d\mu_f(x).$$

Let $x \in \mathbb{B}$ be arbitrary and let $h = f \circ \varphi_x$. By Corollary 7.3.5 the least $\mathcal{H}$-harmonic majorant of f, and hence also of h, is the zero function. Thus h satisfies (9.1.1), and therefore

$$f(x) = h(0) = -\int_{\mathbb{B}} g_h(y)d\mu_h(y)$$

where μ_h is the Riesz measure of h. But if $\psi \in C_c^2(\mathbb{B})$, by the invariance of Δ_h and τ,

$$\begin{aligned} \int_{\mathbb{B}} \psi d\mu_h &= \int_{\mathbb{B}} (f \circ \varphi_x)\Delta_h \psi d\tau \\ &= \int_{\mathbb{B}} f\Delta_h(\psi \circ \varphi_x)d\tau = \int_{\mathbb{B}} (\psi \circ \varphi_x)d\mu_f. \end{aligned}$$

As a consequence,

$$\int_{\mathbb{B}} \psi \, d\mu_h = \int_{\mathbb{B}} (\psi \circ \varphi_x) d\mu_f$$

for any non-negative Borel measurable function ψ. Therefore

$$f(x) = -\int_{\mathbb{B}} g_h(\varphi_x(y)) d\mu_f(y) = -\int_{\mathbb{B}} G_h(x,y) d\mu_f(y),$$

which proves the result. □

Theorem 9.1.3 (Riesz Decomposition Theorem) *Suppose f is $\mathcal{H}$-subharmonic on $\mathbb{B}$ and f has an $\mathcal{H}$-harmonic majorant. Then*

$$f(x) = F_f(x) - \int_{\mathbb{B}} G_h(x,y) d\mu_f(y),$$

where μ_f is the Riesz measure of f and F_f is the least $\mathcal{H}$-harmonic majorant of f.

Proof. Let F_f be the least $\mathcal{H}$-harmonic majorant of f and let $h(x) = f(x) - F_f(x)$. Then h satisfies (9.1.2). Furthermore, since F_f is $\mathcal{H}$-harmonic, $\mu_h = \mu_f$, which proves the result. □

9.2 Applications of the Riesz Decomposition Theorem

In this section we prove several consequences of the Riesz decomposition theorem. We begin with the following lemma, which in itself is an application of Theorem 9.1.3.

Lemma 9.2.1 *For all $y \in \mathbb{B}$,*

$$\int_{\mathbb{B}} G_h(x,y) d\nu(x) = \frac{1}{2n(n-1)}(1-|y|^2)^{n-1}.$$

Proof. Set $\psi(x) = 1 - (1-|x|^2)^{n-1}$. Using the radial form of Δ_h we have

$$\Delta_h \psi(x) = 2n(n-1)(1-|x|^2)^n.$$

Therefore,

$$\int_{\mathbb{B}} G_h(x,y) \Delta_h \psi(x) d\tau(x) = 2n(n-1) \int_{\mathbb{B}} G_h(x,y) d\nu(x).$$

But ψ is a non-negative C^2 $\mathcal{H}$-subharmonic function with least $\mathcal{H}$-harmonic majorant $F_\psi(x) = 1$. Thus by Theorem 9.1.3,

$$\int_{\mathbb{B}} G_h(x,y)\Delta_h\psi(x)d\tau(x) = F_\psi(y) - \psi(y) = (1-|y|^2)^{n-1},$$

from which the result now follows. □

Theorem 9.2.2 *Let f be a non-negative $\mathcal{H}$-subharmonic function with least $\mathcal{H}$-harmonic majorant F_f and Riesz measure μ_f. Then*

$$F_f(0) = \int_{\mathbb{B}} f(y)\,d\nu(y) + \frac{1}{2n(n-1)}\int_{\mathbb{B}}(1-|y|^2)^{n-1}d\mu_f(y).$$

Proof. Since F_f is $\mathcal{H}$-harmonic,

$$F_f(0) = \int_{\mathbb{S}} F_f(rt)d\sigma(t).$$

Multiplying the above by nr^{n-1} and integrating from 0 to 1 gives

$$F_f(0) = \int_{\mathbb{B}} F_f(x)d\nu(x),$$

which by Theorem 9.1.3 and Fubini's theorem

$$= \int_{\mathbb{B}} f(x)d\nu(x) + \int_{\mathbb{B}}\left[\int_{\mathbb{B}} G_h(x,y)d\nu(x)\right]d\mu_f(y).$$

The result now follows by Lemma 9.2.1 □

Theorem 9.2.3 *Let $f \not\equiv -\infty$ be an $\mathcal{H}$-subharmonic function on $\mathbb{B}$ with Riesz measure μ_f. Then the following are equivalent.*

(a) *f has an $\mathcal{H}$-harmonic majorant on $\mathbb{B}$.*

(b) $\displaystyle\int_{\mathbb{B}} G_h(y_o,x)d\mu_f(x) < \infty$ *for some $y_o \in \mathbb{B}$.*

(c) $\displaystyle\int_{\mathbb{B}}(1-|x|^2)^{n-1}d\mu_f(x) < \infty.$

Proof. If (a) holds, then by the Riesz decomposition theorem

$$f(y) = H_f(y) - \int_{\mathbb{B}} G(y,x)d\mu_f(x).$$

By Theorem 4.4.3 $f(y) > -\infty$ a.e. on $\mathbb{B}$. Thus $\int_{\mathbb{B}} G_h(y,x)d\mu_f(x) < \infty$ for almost every $y \in \mathbb{B}$.

Suppose now that (b) holds for some $y_o \in \mathbb{B}$. By (3.2.4),

$$G_h(y_o,x) \geq c_n(1-|\varphi_{y_o}(x)|)^{n-1} = c_n\frac{(1-|y_o|^2)^{n-1}(1-|x|^2)^{n-1}}{\rho(y_o,x)}.$$

But by Exercise 2.4.1, $\rho(y_o, x) \le (1 + |y_o||x|)^2 \le 4$. Therefore

$$G_h(y_o, x) \ge c_n(1 - |x|^2)^{n-1}(1 - |y_o|^2)^{n-1},$$

from which (c) follows.

Finally, suppose that (c) holds. Fix c, $0 < c < 1$, and define V_1 and V_2 on $\mathbb{B}$ by

$$V_1(x) = \int_{B_c} G_h(y, x)d\mu_f(x),$$

$$V_2(x) = \int_{\mathbb{B}\setminus B_c} G_h(y, x)d\mu_f(x).$$

Since $\mu_{f|_{B_c}}$ is a finite measure on $\mathbb{B}$, V_1 is $\mathcal{H}$-superharmonic on $\mathbb{B}$. Also, by (3.2.3),

$$G_h(0, x) \le \frac{1}{nc^{n-2}}(1 - |x|^2)^{n-1} \quad \text{for all } x \in \mathbb{B} \setminus B_c.$$

Therefore, $V_2(0) < \infty$ and thus V_2 is $\mathcal{H}$-superharmonic on $\mathbb{B}$. Hence $V_f = V_1 + V_2$ is $\mathcal{H}$-superharmonic on $\mathbb{B}$.

Set $h(x) = f(x) + V_f(x)$, which is defined a.e. on $\mathbb{B}$ and is locally integrable. Let $\psi \in C_c^2(\mathbb{B})$, $\psi \ge 0$, and consider $\int h\Delta_h\psi d\tau$. By (9.1.1)

$$\int_{\mathbb{B}} G_h(y, x)\Delta_h\psi(x)d\tau(x) = -\psi(y).$$

Thus by Fubini's theorem and the definition of μ_f,

$$\int_{\mathbb{B}} V_f\Delta_h\psi d\tau = -\int_{\mathbb{B}} \psi d\mu_f = -\int_{\mathbb{B}} f\Delta_h\psi d\tau.$$

Therefore,

$$\int_{\mathbb{B}} h(x)\Delta_h\psi(x)d\tau(x) = 0$$

for all $\psi \in C_c^2(\mathbb{B})$ with $\psi \ge 0$. Thus by Theorem 4.6.2 there exists an $\mathcal{H}$-harmonic function H such that $h(x) = H(x)$ a.e. on $\mathbb{B}$. Since $V_f \ge 0$ we have $f(x) \le H(x)$ a.e. However, as a consequence of Theorem 4.3.5, if $a \in \mathbb{B}$ and $r > 0$ is sufficiently small,

$$\begin{aligned} f(a) &\le \frac{1}{\tau(E(a,r))}\int_{E(a,r)} f(x)d\tau(x) \\ &\le \frac{1}{\tau(E(a,r))}\int_{E(a,r)} H(x)d\tau(x) = H(a). \end{aligned}$$

Therefore $f(a) \le H(a)$ for all $a \in \mathbb{B}$, which proves the result. □

Theorem 9.2.4 *Let f be a non-negative $\mathcal{H}$-subharmonic function on $\mathbb{B}$. Then*

(a) *$f \in \mathcal{S}^p$, $p \geq 1$, if and only if $\int_{\mathbb{B}} (1-|y|^2)^{n-1} d\mu_{f^p}(y) < \infty$, where μ_{f^p} is the Riesz measure of f^p.*

(b) *If this is the case, then*

$$\|f\|_p^p = f^p(0) + \int_{\mathbb{B}} g_h(y) d\mu_{f^p}(y). \tag{9.2.1}$$

Proof. The result (a) is just a restatement of Theorem 9.2.3.

(b) Suppose $f \in \mathcal{S}^p, p \geq 1$. Then the least $\mathcal{H}$-harmonic majorant H_{f^p} of f^p is given by

$$H_{f^p}(x) = \int_{\mathbb{S}} P_h(x,t)\hat{f}^p(t)d\sigma(t), \quad p > 1,$$

where $\hat{f} \in L^p(\mathbb{S})$ is the boundary function of f with $\|f\|_p = \|\hat{f}\|_p$. When $p = 1$, H_f is the Poisson integral of a measure ν_f where ν_f is the boundary measure of f. In this case, $\|f\|_1 = \nu_f(\mathbb{S}) = H_f(0)$. By the Riesz decomposition theorem,

$$H_{f^p}(x) = f^p(x) + \int_{\mathbb{B}} G_h(x,y) d\mu_{f^p}(y). \tag{9.2.2}$$

Hence

$$\|f\|_p^p = H_{f^p}(0) = f^p(0) + \int_{\mathbb{B}} g_h(y) d\mu_{f^p}(y) \tag{9.2.3}$$

which proves the result. □

Remarks 9.2.5 **(a)** *Equation (9.2.1) is sometimes referred to as the* **Hardy–Stein identity**.

(b) *Although we assumed $p \geq 1$ in the previous theorem, the conclusion is still valid for $0<p<1$ provided we assume in addition that f^p is $\mathcal{H}$-subharmonic. In this case $f^p(x) \leq P_h[\nu_{f^p}](x)$ for some non-negative measure ν_{f^p} on $\mathbb{S}$ with $\|f\|_p^p = \nu_{f^p}(\mathbb{S}) = H_{f^p}(0)$.*

(c) *Since $g_h(y) \geq c_n(1-|y|^2)^{n-1}$ we have*

$$\|f\|_p^p \geq f^p(0) + c_n \int_{\mathbb{B}} (1-|y|^2)^{n-1} d\mu_{f^p}(y).$$

In the other direction, integrating (9.2.2) over B_{r_o}, $0 < r_o < \frac{1}{2}$, gives

$$H_{f^p}(0) = \frac{1}{r_o^n}\left[\int_{B_{r_o}} f^p(x)d\nu(x) + \int_{\mathbb{B}}\int_{B_{r_o}} G_h(x,y)d\nu(x)d\mu_{f^p}(y)\right],$$

which by Lemma 9.2.1

$$\leq \frac{1}{r_o^n}\left[\int_{B_{r_o}} f^p(x) + C_n \int_{\mathbb{B}} (1-|y|^2)^{n-1} d\mu_{f^p}(y)\right].$$

Suppose in addition that f^p *is* C^2 *and* $\Delta_h f^p$ *has quasi-nearly* $\mathcal{H}$*-subharmonic behavior. Then by Theorem 4.1.1, for* $0 < r < r_o$,

$$\int_{\mathbb{S}} f^p(r\zeta)d\sigma(\zeta) \leq f^p(0) + C_n r_o^2 \sup_{x\in B_{r_o}} \Delta_h f^p(x).$$

From this it now follows that

$$\|f\|_p^p \leq f^p(0) + C_{n,r_o}\int_{\mathbb{B}} (1-|y|^2)^{n-1}\Delta_h f^p(y)d\tau(y).$$

Corollary 9.2.6 *Let* h *be* $\mathcal{H}$*-harmonic on* $\mathbb{B}$*. If* $p \geq 2$*, or if* h *is non-zero when* $1 < p < 2$*, then* $h \in \mathcal{H}^p$ *if and only if*

$$\int_{\mathbb{B}} (1-|x|^2)^{n-1}|h(x)|^{p-2}|\nabla^h h(x)|^2 d\tau(x) < \infty.$$

Proof. If h satisfies the hypothesis, then $|h|^p \in C^2(\mathbb{B})$, $1 < p < \infty$, with

$$\Delta_h|h|^p = p(p-1)|h|^{p-2}|\nabla^h h|^2.$$

The conclusion now follows by the previous theorem. □

A similar result also holds for $\mathcal{H}_\varphi$, where φ is strongly convex on $(-\infty,\infty)$. In this case, if h is non-zero on $\mathbb{B}$ and φ is C^2, then by Exercise 3.5.3,

$$\Delta_h\varphi(|h|) = \varphi''(|h|)|\nabla^h h|^2.$$

Definition 9.2.7 *If* μ *is a regular Borel measure on* $\mathbb{B}$*, the function* G_μ *defined by*

$$G_\mu(x) = \int_{\mathbb{B}} G_h(x,y)\,d\mu(y)$$

is called the (invariant) **Green potential** *of* μ *provided* $G_\mu \not\equiv +\infty$.

If $G_\mu \not\equiv +\infty$, then since $x \to G_h(x,y)$ is $\mathcal{H}$-superharmonic, by Tonelli's theorem so is the function G_μ. Furthermore, by Exercise 4.8.7, G_μ is $\mathcal{H}$-harmonic on $\mathbb{B}\setminus \operatorname{supp}\mu$. Also, as in the proof of Theorem 9.2.3 it is easy to show that $G_\mu \not\equiv +\infty$ if and only if

$$\int_{\mathbb{B}} (1-|x|^2)^{n-1} d\mu(x) < \infty. \tag{9.2.4}$$

Theorem 9.2.8 *A non-negative $\mathcal{H}$-superharmonic function V is a Green potential on $\mathbb{B}$ if and only if*

$$\lim_{r\to 1}\int_{\mathbb{S}} V(rt)d\sigma(t) = 0. \tag{9.2.5}$$

Proof. If V satisfies (9.2.5), then by the Riesz decomposition theorem the function V is the Green potential of its Riesz measure.

Conversely, suppose $V = G_\mu$ where μ satisfies (9.2.4). Without loss of generality we assume $n > 2$. Fix δ, $0 < \delta < \frac{1}{2}$. By inequality (3.2.3),

$$G_h(0,x) \le C_\delta(1-|x|^2)^{n-1}$$

for all x, $|x| \ge \delta$. Therefore, since $V(x) < \infty$ a.e., by (9.2.4)

$$\int_{\mathbb{B}\setminus B_\delta} G_h(0,x)d\mu(x) < \infty.$$

Let $\epsilon > 0$ be given. Choose R, $\delta < R < 1$, such that

$$\int_{\mathbb{B}\setminus B_R} G_h(0,x)d\mu(x) < \epsilon.$$

For this R write

$$\begin{aligned} V(y) &= \int_{B_R} G_h(y,x)d\mu(x) + \int_{\mathbb{B}\setminus B_R} G_h(y,x)d\mu(x) \\ &= V_1(y) + V_2(y). \end{aligned}$$

Since $y \to G_h(y,x)$ is $\mathcal{H}$-superharmonic, by Tonelli's theorem

$$\begin{aligned} \int_{\mathbb{S}} V_2(rt)d\sigma(t) &= \int_{B\setminus B_R}\int_{\mathbb{S}} G_h(rt,x)d\sigma(t)d\mu(x) \\ &\le \int_{B\setminus B_R} G_h(0,x)d\mu(x) < \epsilon. \end{aligned}$$

Therefore

$$\limsup_{r\to 1}\int_{\mathbb{S}} V_2(rt)d\sigma(t) < \epsilon.$$

Consider $V_1(rt)$. By (3.2.5),

$$\begin{aligned} G_h(rt,x) &\le \frac{(1-r^2)^{n-1}(1-|x|^2)^{n-1}}{n|rt-x|^{n-2}\{|rt-x|^2+(1-r^2)(1-|x|^2)\}^{n/2}} \\ &\le \frac{(1-r^2)^{n-1}(1-|x|^2)^{n-1}}{n|rt-x|^{2n-2}}. \end{aligned}$$

But for $x \in B_R$ and $(1+R)/2 < r < 1$, $|rt - x| \geq r - |x| > (1-R)/2$. Hence

$$G_h(rt, x) \leq \left(\frac{2}{1-R}\right)^{2n-2} \frac{1}{n}(1-r^2)^{n-1}(1-|x|^2)^{n-1},$$

and thus

$$V_1(rt) \leq C_R(1-r^2)^{n-1} \int_{B_R} (1-|x|^2)^{n-1} d\mu(x).$$

Therefore $\lim_{r\to 1} V_1(rt) = 0$ uniformly on $\mathbb{B}$ and thus

$$\limsup_{r\to 1} \int_{\mathbb{S}} V(rt) d\sigma(t) < \epsilon,$$

which proves the result. □

9.3 Integrability of $\mathcal{H}$-Superharmonic Functions

In [8] D. H. Armitage proved that if $V(x) = G^e_\mu(x)$, where G^e is the Euclidean Green's function on $\mathbb{B}$ and μ is a measure on $\mathbb{B}$ satisfying

$$\int_{\mathbb{B}} (1-|x|) d\mu(x) < \infty.$$

Then $V \in L^p(\mathbb{B}, \nu)$ for all $p, 0 < p < n/(n-1)$, with

$$\int_{\mathbb{B}} V^p(x) dx \leq A(n,p) V^p(0).$$

The analogous result for $\mathcal{M}$-superharmonic functions on the unit ball $\mathbb{B}$ in $\mathbb{C}^m$ was proved by S. Zhao in [96]. The result of Zhao states that if V is a non-negative $\mathcal{M}$-superharmonic function on $\mathbb{B} \subset \mathbb{C}^m = \mathbb{R}^{2m}$, then $V \in L^p(\mathbb{B}, \nu)$ for all p, $0 < p < (m+1)/m$. In this case, for $m > 1$, the constant $(m+1)/m$ is strictly greater than $2m/(2m-1)$. Of course, when $m = 1$, the two upper bounds agree.

In this section we prove the analogous result for non-negative $\mathcal{H}$-superharmonic functions on $\mathbb{B} \subset \mathbb{R}^n$. As we will see, our upper bounds are similar to those obtained for Euclidean superharmonic functions. Our main result of this section is the following analogue of Theorem 3 of D. H. Armitage [8]. Our method of proof, with minor modifications, is similar to that used by S. Zhao in [96].

Theorem 9.3.1 *Let U be a non-negative $\mathcal{H}$-superharmonic function on $\mathbb{B}$. Then for each $a \in \mathbb{B}$, there exists a constant $A(n,p,a)$, independent of U, such that*

$$\int_{\mathbb{B}} U^p(x)d\nu(x) \le A(n,p,a)U^p(a)$$

for all p, $0 < p < n/(n-1)$.

We begin by proving several results concerning integrability of non-negative $\mathcal{H}$-harmonic functions.

Lemma 9.3.2 *For all* $\zeta \in \mathbb{S}$, *there exists a finite constant* $A(n,p)$, *such that*

$$\int_{\mathbb{B}} P_h^p(x,\zeta)d\nu(x) \le A(n,p)$$

for all p, $0 < p < n/(n-1)$. *Furthermore, for* $p = n/(n-1)$

$$\int_{\mathbb{B}} P_h^p(x,\zeta)d\nu(x) = +\infty.$$

In the above, ν *denotes normalized Lebesgue measure on* $\mathbb{B}$.

Proof. Since $x \to P_h^p(x,\zeta)$ is $\mathcal{H}$-superharmonic on $\mathbb{B}$ for all $p, 0 < p \le 1$, it suffices to only consider the case $p > 1$. Hence

$$\int_{\mathbb{B}} P_h^p(x,\zeta)d\nu(x) = n\int_0^1 r^{n-1}(1-r^2)^{p(n-1)}\int_{\mathbb{S}} \frac{d\sigma(t)}{|rt-\zeta|^{2p(n-1)}}\,dr,$$

which by Theorem 5.5.7 for $p > 1$

$$\approx \int_0^1 r^{n-1}(1-r^2)^{p(n-1)}(1-r)^{-2p(n-1)+(n-1)}dr$$

$$\approx \int_0^1 r^{n-1}(1-r)^{(n-1)-p(n-1)}dr.$$

However, the above integral is finite if and only if $p < n/(n-1)$. □

Theorem 9.3.3 *Let* $h(x)$ *be a non-negative* $\mathcal{H}$*-harmonic function on* $\mathbb{B}$. *Then*

$$\int_{\mathbb{B}} h^p(x)d\nu(x) \le A(n,p)h^p(0)$$

for all p, $0 < p < n/(n-1)$.

Proof. Again, since h^p is $\mathcal{H}$-superharmonic for all p, $0 < p \le 1$, we only need to consider the result for $p > 1$. By Theorem 7.1.1, $h(x) = P_h[\mu](x)$ where μ is a non-negative measure on $\mathbb{S}$. Therefore

$$\int_{\mathbb{B}} h^p(x)d\nu(x) = \int_{\mathbb{B}}\left(\int_{\mathbb{S}} P_h(x,\zeta)d\mu(\zeta)\right)^p d\nu(x),$$

which by Hölder's inequality and Lemma 9.3.2

$$\leq \int_{\mathbb{B}} \left(\int_{\mathbb{S}} P_h^p(x,\zeta) d\mu(\zeta) \right) \left(\int_{\mathbb{S}} d\mu(\zeta) \right)^{p-1} d\nu(x)$$
$$= h^{p-1}(0) \int_{\mathbb{S}} \left(\int_{\mathbb{B}} P_h^p(x,\zeta) d\nu(x) \right) d\mu(\zeta),$$

which for $p < n/(n-1)$

$$\leq A(n,p) h^{p-1}(0) \int_{\mathbb{S}} d\mu(\zeta) = A(n,p) h^p(0).$$

□

We next turn our attention to the integrability of Green potentials. For the proofs we require the following lemma.

Lemma 9.3.4 *Fix δ, $0 < \delta < 1$. Then*

$$G_h(x,y) \leq \begin{cases} C_\delta\, G_h(0,y) \dfrac{(1-|x|^2)^{n-1}}{\rho(x,y)^{n-1}} & \text{for all } x \in \mathbb{B} \setminus E(y,\delta), \\ C_n\, G_h(0,y) \dfrac{1}{|x-y|^{n-1}} & \text{for all } x \in \mathbb{B}. \end{cases}$$

Proof. For $x \in \mathbb{B} \setminus E(y,\delta)$, by (3.2.4)

$$G_h(x,y) \leq \frac{1}{n\delta^{n-2}} \frac{(1-|y|^2)^{n-1}(1-|x|^2)^{n-1}}{\rho(x,y)^{n-1}}.$$

But by (3.2.3) $(1-|y|^2)^{n-1} \leq c_n G_h(0,y)$, from which the first inequality follows. On the other hand, by (3.2.5),

$$G_h(x,y) \leq \frac{(1-|y|^2)^{n-1}(1-|x|^2)^{n-1}}{n|x-y|^{n-2}\rho(x,y)^{\frac{n}{2}}}.$$

But

$$\rho(x,y)^{\frac{n}{2}} = \rho(x,y)^{\frac{n-1}{2}} \rho(x,y)^{\frac{1}{2}} \geq |x-y| \rho(x,y)^{\frac{n-1}{2}}.$$

Also

$$\rho(x,y) \geq ||x|-|y||^2 + (1-|x|^2)(1-|y|^2) = (1-|x||y|)^2 \geq (1-|x|)^2.$$

Therefore

$$\rho(x,y)^{\frac{n-1}{2}} \geq (1-|x|)^{n-1},$$

and thus

$$G_h(x,y) \leq C_n G_h(0,y) |x-y|^{-(n-1)}.$$

□

Theorem 9.3.5 *For all p, $0 < p < n/(n-1)$,*

$$\int_{\mathbb{B}} G_h^p(x,y)d\nu(x) \le A(n,p)G^p(0,y).$$

Proof. Suppose first that $0 < p \le 1$. Then by Hölder's inequality, Lemma 9.2.1, and inequality (3.2.2),

$$\int_{\mathbb{B}} G_h^p(x,y)d\nu(x) \le \left(\int_{\mathbb{B}} G_h(x,y)d\nu(x)\right)^p$$
$$\le C_n^p\left[(1-|y|^2)^{n-1}\right]^p \le C_n^p G_h^p(0,y).$$

Suppose now that $p > 1$. Set $E(y) = E(y,\delta)$ where $0 < \delta < \frac{1}{2}$, and write

$$\int_{\mathbb{B}} G_h^p(x,y)d\nu(x) = \int_{E(y)} G_h^p(x,y)d\nu(x) + \int_{\mathbb{B}\setminus E(y)} G_h^p(x,y)d\nu(x).$$

By Lemma 9.3.4

$$\int_{\mathbb{B}\setminus E(y)} G_h^p(x,y)d\nu(x) \le C_{\delta,n}G_h^p(0,y)\int_0^1 r^{n-1}\int_{\mathbb{S}}\frac{(1-r^2)^{p(n-1)}}{\rho(rt,y)^{p(n-1)}}\,d\sigma(t)\,dr.$$

Write $y = s\eta$, $\eta \in \mathbb{S}$. Since

$$\rho(rt,s\eta) = \rho(t,rs\eta) = |t-rs\eta|^2,$$

we have by Theorem 5.5.7

$$\int_{\mathbb{S}}\frac{d\sigma(t)}{\rho(rt,y)^{p(n-1)}} = \int_{\mathbb{S}}\frac{d\sigma(t)}{|t-rs\eta|^{2p(n-1)}} \le C(1-rs)^{-2p(n-1)+(n-1)}$$
$$\le C(1-r)^{-2p(n-1)+(n-1)}.$$

Therefore

$$\int_{\mathbb{B}\setminus E(y)} G_h^p(x,y)d\nu(x) \le C_{\delta,n}G_h^p(0,y)\int_0^1 r^{n-1}(1-r)^{(n-1)-p(n-1)}dr.$$

The above integral, however, is finite if and only if $p < n/(n-1)$.

Suppose $x \in E(y)$. Since $E(y,\delta)$ is also a Euclidean ball with center c_y and radius ρ_y given by

$$c_y = \frac{(1-\delta^2)y}{1-|y|^2\delta^2} \quad \text{and} \quad \rho_y = \frac{\delta(1-|y|^2)}{1-|y|^2\delta^2},$$

we have for $0 < \delta < \frac{1}{2}$ that

$$|x| < |c_y| + |x-c_y| = \frac{|y|+\delta}{1+|y|\delta}.$$

Therefore $E(y,\delta) \subset B(0,\rho)$ with $\rho = (|y|+\delta)/(1+|y|\delta)$. Hence by Lemma 9.3.4

$$\int_{E(y)} G_h^p(x,y)d\nu(x) \le C\, G_h^p(0,y) \int_{B(0,\rho)} \frac{d\nu(x)}{|x-y|^{p(n-1)}}.$$

But by the change of variables $w = x - y$ and integration in polar coordinates,

$$\int_{B(0,\rho)} \frac{d\nu(x)}{|x-y|^{p(n-1)}} \le \int_{B(0,2)} \frac{d\nu(w)}{|w|^{p(n-1)}} = n \int_0^2 r^{n-1-p(n-1)},$$

which is finite provided $p < n/(n-1)$. □

As an immediate consequence of the previous theorem, we have the following corollary.

Corollary 9.3.6 *Let $f \in L^q(\mathbb{B},\nu)$ for some $q > n$, and let*

$$V_f(x) = \int_{\mathbb{B}} f(y)G_h(x,y)d\nu(y). \tag{9.3.1}$$

Then

$$\lim_{|x|\to 1} V_f(x) = 0.$$

Proof. Let p denote the conjugate exponent of n. Since $q > n$ we have $1 < p < n/(n-1)$. Hence by Hölder's inequality,

$$\begin{aligned} |V_f(x)| &\le \|f\|_q \left(\int_{\mathbb{B}} G_h^p(x,y)d\nu(y)\right)^{1/p} \\ &\le A(n,p)\|f\|_q G_h(0,x), \end{aligned}$$

from which the result now follows. □

Theorem 9.3.7 *Let G_μ be the potential of a measure μ satisfying (9.2.4). Then*

$$\int_{\mathbb{B}} G_\mu^p(y)d\nu(y) \le A(n,p)G_\mu^p(0)$$

for all p, $0 < p < n/(n-1)$.

Proof. The theorem is certainly true if $G_\mu(0) = +\infty$. Hence, assume that $G_\mu(0) < \infty$. For $0 < p \le 1$ we have

$$\begin{aligned} \int_{\mathbb{B}} G_\mu^p(y)d\nu(y) &\le \left(\int_{\mathbb{B}} G_\mu(y)d\nu(y)\right)^p \\ &= \left(n\int_0^1 r^{n-1}\int_{\mathbb{S}} G_\mu(rt)d\sigma(t)dr\right)^p \le G_\mu^p(0). \end{aligned}$$

The last inequality follows since G_μ is $\mathcal{H}$-superharmonic.

For $p > 1$, by the continuous version of Minowski's inequality [18, VI.11.13],

$$\int_{\mathbb{B}} G_\mu^p(x)d\nu(x) = \int_{\mathbb{B}} \left(\int_{\mathbb{B}} G_h(x,y)d\mu(y) \right)^p d\nu(x),$$
$$\leq \int_{\mathbb{B}} \int_{\mathbb{B}} G_h^p(x,y)d\mu(x)\, d\nu(y),$$

which by the previous theorem

$$\leq A(n,p) \int_{\mathbb{B}} G_h^p(0,y)d\mu(y) = A(n,p)G_\mu^p(0).$$

□

Proof of Theorem 9.3.1 Fix $a \in \mathbb{B}$ and let $V = U \circ \varphi_a$. Then V is $\mathcal{H}$-superharmonic and

$$\int_{\mathbb{B}} V^p(x)d\nu(x) = \int_{\mathbb{B}} (1-|x|^2)^n U^p(\varphi_a(x))\, d\tau(x),$$

which by the invariance of τ

$$= \int_{\mathbb{B}} (1-|\varphi_a(x)|^2)^n U^p(x)\, d\tau(x).$$

But

$$1 - |\varphi_a(x)|^2 = \frac{(1-|x|^2)(1-|a|^2)}{\rho(x,a)} \geq \frac{(1-|x|^2)(1-|a|^2)}{(1+|a|)^2}.$$

Hence

$$\int_{\mathbb{B}} U^p(x)d\nu(x) \leq \left(\frac{1+|a|}{1-|a|} \right)^n \int_{\mathbb{B}} V^p(x)d\nu(x).$$

Since V is a non-negative $\mathcal{H}$-superharmonic function, by the Riesz decomposition theorem

$$V(x) = h(x) + G_\mu(x),$$

where h is $\mathcal{H}$-harmonic and G_μ is the Green potential of a measure μ. Thus by Theorems 9.3.3 and 9.3.7

$$\int_{\mathbb{B}} V^p(x)d\nu(x) \leq A(n,p)\left\{ h^p(0) + G_\mu^p(0) \right\}$$
$$\leq A(n,p)V^p(0) = A(n,p)U^p(a).$$

□

9.4 Boundary Limits of Green Potentials

In Corollary 9.3.6 we proved that if $f \in L^q(\mathbb{B}, \nu)$ for some $q > n$, then

$$\lim_{|x|\to 1} Gf(x) = 0,$$

where Gf was given by (9.3.1). Also, it is easily shown that if μ has compact support K, then

$$G_\mu(x) \le C_K(1 - |x|^2)^{n-1} \int_K (1 - |y|^2)^{n-1} d\mu(y),$$

and thus $\lim_{|x|\to 1} G_\mu(x) = 0$ uniformly.

Our main result of this section is the following theorem concerning radial limits of Green potentials.

Theorem 9.4.1 *Let G_μ be the Green potential of a measure μ satisfying (9.2.4). Then*

$$\lim_{r\to 1} G_\mu(rt) = 0 \qquad \text{for almost every } t \in \mathbb{S}.$$

We will also give an example of a measure μ satisfying (9.2.4) for which

$$\lim_{\substack{x\to\zeta \\ x\in\Gamma_\alpha(\zeta)}} G_\mu(x) = +\infty \qquad \text{for almost every } t \in \mathbb{S}$$

for every $\alpha > 1$.

Our method of proof of Theorem 9.4.1 will follow the technique used by D. Ullrich in proving the analogous result for invariant potentials on the unit ball in $\mathbb{C}^m$ [93], [94]. (See also [84, Theorem 8.1].) For a regular Borel measure μ on $\mathbb{B}$ satisfying (9.2.4) define the functions V_1 and V_2 on $\mathbb{B}$ by

$$V_1(x) = \int_{E(x)} G_h(x, y)d\mu(y), \quad \text{and}$$

$$V_2(x) = \int_{\mathbb{B}\setminus E(x)} G_h(x, y)d\mu(y),$$

where $E(x) = E(x, \frac{1}{2})$. In our first theorem we prove that the function V_2 has non-tangential limit zero at almost every point of $\mathbb{S}$.

Theorem 9.4.2 *Let μ and V_2 be as above. Then*

$$\lim_{\substack{x\to\zeta \\ x\in\Gamma_\alpha(\zeta)}} V_2(x) = 0 \qquad \text{for almost every } \zeta \in \mathbb{S}.$$

Proof. For $y \in \mathbb{B} \setminus E(x)$ we have

$$V_2(x) \le \int_{\mathbb{B}} \frac{(1 - |x|^2)^{n-1}(1 - |y|^2)^{n-1}}{\rho(x, y)^{n-1}} d\mu(y).$$

Fix an R, $0 < R < 1$. Since

$$\rho(x,y) = |x-y|^2 + (1-|x|^2)(1-|y|^2) \ge (1-|x||y|)^2,$$

for $y \in B_R$ we have $\rho(x,y) \ge (1-R)^2$. Therefore,

$$\begin{aligned}\int_{B_R} \frac{(1-|x|^2)^{n-1}(1-|y|^2)^{n-1}}{\rho(x,y)^{n-1}} d\mu(y) &\le \frac{(1-|x|^2)^{n-1}}{(1-R)^{2(n-1)}} \int_{B_R} (1-|y|^2)^{n-1} d\mu(y)\\ &= C_R(1-|x|^2)^{n-1}.\end{aligned}$$

Let $A_R = \{y : R < |y| < 1\}$. Define a measure ν_R on $\mathbb{S}$ as follows: for $h \in C(\mathbb{S})$,

$$\int_{\mathbb{S}} h(t)d\nu_R(t) = \int_{A_R} h\left(\tfrac{y}{|y|}\right)(1-|y|^2)^{n-1} d\mu(y).$$

Then ν_R is a finite Borel measure on $\mathbb{S}$. For $y \in A_R$, write $y = r\eta$, $\eta \in \mathbb{S}$. Then

$$\rho(x, r\eta) = |rx - \eta|^2 \ge \tfrac{1}{4}|x-\eta|^2.$$

Therefore

$$\begin{aligned}\int_{A_R} \frac{(1-|x|^2)^{n-1}(1-|y|^2)^{n-1}}{\rho(x,y)} d\mu(y) &\le \int_{A_R} \frac{(1-|x|^2)^{n-1}}{\left|x - \frac{y}{|y|}\right|^{2(n-1)}} d\mu(y)\\ &= C_n \int_{\mathbb{S}} P_h(x,t)d\nu_R(t),\end{aligned}$$

where C_n is a constant depending only on n. Therefore

$$V_2(x) \le C_R(1-|x|^2)^{n-1} + C_n P_h[\nu_R](x),$$

and thus

$$\begin{aligned}\limsup_{\substack{x\to\zeta\\ x\in\Gamma_\alpha(\zeta)}} V_2(x) &\le C_n \limsup_{\substack{x\to\zeta\\ x\in\Gamma_\alpha(\zeta)}} P_h[\nu_R]x\\ &\le \sup_{x\in\Gamma_\alpha(\zeta)} C_n P_h[\nu_R](x),\end{aligned}$$

which by Theorem 8.2.3

$$\le C_n A_\alpha M[\nu_R](\zeta),$$

where A_α is a constant depending only on α. But by Theorem 8.1.2 there exists a constant B_n, depending only n, such that

$$\sigma(\{\zeta \in \mathbb{S} : M[\nu_R](\zeta) > \lambda\}) < \frac{B_n}{\lambda}\nu_R(\mathbb{S}).$$

Therefore, there exists a constant $C_{n,\alpha}$, depending only on α and n, such that

$$\sigma(E_\lambda) \le \frac{C_{n,\alpha}}{\lambda} \nu_R(\mathbb{S}),$$

where

$$E_\lambda = \{\zeta \in \mathbb{S} : \limsup_{\substack{x \to \zeta \\ x \in \Gamma_\alpha(\zeta)}} V_2(x) > \lambda\}.$$

Suppose

$$\limsup_{\substack{x \to \zeta \\ x \in \Gamma_\alpha(\zeta)}} V_2(x) > 0$$

on a set of positive measure E. Thus $E \subset \cup_{\lambda>0} E_\lambda$. Suppose there exists a $\lambda > 0$ such that $\sigma(E_\lambda) > 0$. For this λ, choose R, $0 < R < 1$, such that $C_{n,\alpha} \nu_R(\mathbb{S}) < \frac{1}{2}\lambda\sigma(E_\lambda)$ to obtain a contradiction. □

For the proof that $V_1(x)$ has radial limit zero at almost every point of $\mathbb{S}$ we require some preliminary notation and lemmas. As in (2.2.5), for $0 < \delta < 1$, $x \in \mathbb{B}$, set

$$E_\delta(x) = E(x,\delta) = \{y \in \mathbb{B} : |\varphi_x(y)| < \delta\}.$$

Also, for $x \in \mathbb{B}$, $x \neq 0$, let $\Pi(x)$ denote the projection of x onto $\mathbb{S}$, that is,

$$\Pi(x) = \frac{x}{|x|}.$$

For δ, $r \in (0,1)$, set

$$V_\delta^r(\zeta) = \Pi(E_\delta(r\zeta)). \tag{9.4.1}$$

Lemma 9.4.3 *For $\zeta \in \mathbb{S}$, $0 < \delta \le \frac{1}{2}$, $\frac{3}{4} < r < 1$,*

$$\sigma(V_\delta^r(\zeta)) \approx (1-r^2)^{n-1}\delta^{n-1}.$$

Proof. By invariance under the orthogonal group $O(n)$ we can without loss of generality take $\zeta = e_1 = (1,0,\ldots,0)$. For $0 < \delta \le \frac{1}{2}$ and ρ small, let

$$\Omega_\delta^\rho = \{t : 0 < 1-t < \rho^2\delta^2\}$$
$$N_\delta^\rho = \{\zeta \in \mathbb{S} : \zeta_1 \in \Omega_\delta^\rho\}.$$

We will show below that there exist positive constants c, c', independent of δ and ρ, such that

$$N_{c\delta}^\rho \subset V_\delta^r(e_1) \subset N_{c'\delta}^\rho \quad \text{where } \rho^2 = 1 - r^2. \tag{9.4.2}$$

Assuming that (9.4.2) holds we then obtain that $\sigma(V_\delta^r(e_1)) \approx \sigma(N_\delta^\rho)$. But for $n \geq 3$, by (5.5.11) and Exercise 8.7.1,

$$\sigma(N_\delta^\rho) = c_n \int_{\Omega_\delta^\rho} (1-s^2)^{\frac{n-3}{2}}\,ds$$
$$\approx \delta^{n-1}\rho^{n-1}.$$

To prove (9.4.2) we first show that for an appropriate choice of c, if $\zeta \in N_{c\delta}^\rho$ then $r\zeta \in E_\delta(re_1)$. By identity (2.1.11), $r\zeta \in E_\delta(re_1)$ if and only if

$$\frac{(1-r^2)^2}{\rho(re_1, r\zeta)} = 1 - |\varphi_{re_1}(r\zeta)|^2 > 1-\delta^2,$$

or

$$|1 - r^2\zeta_1|^2 = \rho(re_1, r\zeta) < \frac{1}{1-\delta^2}(1-r^2)^2.$$

But

$$|1-r^2\zeta_1|^2 = 1 - 2r^2\zeta_1 + r^4\zeta_1^2 = (r^2(1-\zeta_1) + (1-r^2))^2.$$

Since $\zeta \in N_{c\delta}^\rho$, with $\rho^2 = (1-r^2)$, we have $1-\zeta_1 < c^2\delta^2(1-r^2)$. Therefore

$$|1-r^2\zeta_1^2| < (c^2\delta^2 r^2(1-r^2) + (1-r^2))^2$$
$$< (1-r^2)^2(1+c^2\delta^2)^2,$$

which if $2c^2 \leq 1$

$$< (1-r^2)^2(1+\delta^2+\delta^4)$$
$$< \frac{1}{1-\delta^2}(1-r^2)^2.$$

Hence for $c^2 \leq \frac{1}{2}$ we have $N_{c\delta}^\rho \subset V_\delta^r(e_1)$.

Suppose $\zeta = (\zeta_1, \zeta') \in V_\delta^r(e_1)$. Then $\zeta = x/|x|$ for some $x \in E_\delta(re_1)$ for $0 < \delta \leq \frac{1}{2}$ and $r \geq \frac{3}{4}$. By Theorem 2.2.2 x is in the Euclidean ball $B(c_{re_1}, \rho_r)$ where

$$c_{re_1} = \frac{(1-\delta^2)re_1}{1-\delta^2 r^2} \quad \text{and} \quad \rho_r = \frac{\delta(1-r^2)}{1-\delta^2 r^2}.$$

Set $x = (x_1, x')$. Then

$$|x - c_{re_1}|^2 = |x_1 - c_{re_1}|^2 + |x'|^2 < \frac{\delta^2(1-r^2)^2}{(1-\delta^2 r^2)^2}.$$

But for $0 < \delta \leq \frac{1}{2}$, $(1-\delta^2 r^2)^2 \geq \frac{1}{4}(1-r^2)$. Therefore

$$|x'|^2 < (2\delta)^2(1-r^2).$$

Also, since $|x_1 - c_{re_1}| < \rho_r$, we have for $0 < \delta \le \frac{1}{2}$ and $r \ge \frac{3}{4}$

$$|x_1| > |c_{re_1}| - \rho_r = \frac{r-\delta}{1-\delta r} \ge c_1 > 0.$$

Therefore we also have $|x| \ge c_1$. Hence, since $x = |x|\zeta$,

$$1 - \zeta_1 = 1 - \frac{x_1}{|x|} \le 1 - \frac{x_1^2}{|x|^2} = \frac{|x'|^2}{|x|^2} < (c'\delta)^2(1-r^2)$$

for an appropriate choice of c'. Therefore $\zeta \in N^{\rho}_{c'\delta}$. □

Definition 9.4.4 *For μ a non-negative measure on $\mathbb{B}$, $0 < \delta < \frac{1}{3}$, $\zeta \in \mathbb{S}$, set*

$$M_\delta\mu(\zeta) = \sup_{0<r<1} \mu(E_\delta(r\zeta)). \tag{9.4.3}$$

Lemma 9.4.5 *Let μ be a regular Borel measure on $\mathbb{B}$ satisfying (9.2.4), and let $\lambda > 0$. Then there exists a constant C, independent of δ, λ, and μ, such that*

$$\sigma(\{\zeta \in \mathbb{S} : M_\delta\mu(\zeta) > \lambda\}) \le C\frac{\delta^{n-1}}{\lambda}\int_{\mathbb{B}} (1-|x|^2)^{n-1} d\mu(x).$$

Proof. Define the finite measure μ^* on $\mathbb{B}$ by

$$d\mu^*(x) = (1-|x|^2)^{n-1} d\mu(x).$$

Fix $\lambda > 0$ and let

$$E_\lambda = \{x \in \mathbb{B} : \mu(E_\delta(x)) > \lambda\}.$$

If $y \in E_\delta(x)$, $0 < \delta < \frac{1}{3}$, then $(1-|y|^2) > \frac{1}{2}(1-|x|^2)$. Therefore, for all $x \in E_\lambda$,

$$\mu^*(E_\delta(x)) > c(1-|x|^2)^{n-1}\lambda,$$

where c is a constant independent of δ. Since $x \to \mu(E_\delta(x))$ is lower semicontinuous, E_λ is an open subset of $\mathbb{B}$. As in Lemma 8.1.3 there exists a countable collection of points $\{x_i\} \subset E_\lambda$ such that $\{E_\delta(x_i)\}$ is pairwise disjoint and

$$E_\lambda \subset \bigcup_i E_{3\delta}(x_i).$$

Set $\zeta_i = x_i/|x_i|$ and $r_i = |x_i|$. Also, let $V_i = V^{r_i}_{3\delta}(\zeta_i)$. Suppose $M_\delta\mu(\zeta) > \lambda$. Then there exists a ρ such that $\rho\zeta \in E_\lambda$. But then $\rho\zeta \in E_{3\delta}(x_i)$ for some i. Therefore $\zeta \in V_i$ and thus

$$\{\zeta \in \mathbb{S} : M_\delta\mu(\zeta) > \lambda\} \subset \bigcup_i V_i.$$

Therefore,

$$\begin{aligned}\sigma(\{\zeta \in \mathbb{S} : M_\delta\mu(\zeta) > \lambda\}) &\le \sum_i \sigma(V_i) \le C\delta^{n-1}\sum_i (1 - r_i^2)^{n-1} \\ &\le C\frac{\delta^{n-1}}{\lambda}\sum \mu^*(E_\delta(x_i)) \\ &\le C\frac{\delta^{n-1}}{\lambda}\int_{\mathbb{B}} (1 - |x|^2)^{n-1} d\mu(x).\end{aligned}$$

□

Lemma 9.4.6 *Let μ be a non-negative regular Borel measure on $\mathbb{B}$ satisfying (9.2.4), and let $\lambda > 0$. Then there exists a positive constant C, independent of λ and μ, such that*

$$\sigma(\{\zeta \in \mathbb{S} : (M_{rad}V_1)(\zeta) > \lambda\}) \le \frac{C}{\lambda}\int_{\mathbb{B}} (1 - |x|^2)^{n-1} d\mu(x),$$

where

$$V_1(x) = \int_{E(x,\frac{1}{2})} G_h(x, y) d\mu(y).$$

Proof. By inequality (3.2.3),

$$V_1(\rho\zeta) \le \frac{1}{n}\int_{E(\rho\zeta)} |\varphi_x(\rho\zeta)|^{2-n} d\mu(x),$$

where $E(\rho\zeta) = E(\rho\zeta, \frac{1}{2})$. Set $\delta_j = 2^{-j/(n-2)}$. Then

$$E(\rho\zeta) = \bigcup_{j=n-2}^{\infty} E_{\delta_j}(\rho\zeta) \setminus E_{\delta_{j+1}}(\rho\zeta).$$

If $x \in E_{\delta_j}(\rho\zeta) \setminus E_{\delta_{j+1}}(\rho\zeta)$, then

$$|\varphi_x(\rho\zeta)|^{n-2} \ge \delta_{j+1}^{n-2} = 2^{-(j+1)}.$$

Therefore,

$$\begin{aligned}V_1(\rho\zeta) &\le C_n \sum_{j=n-2}^{\infty} 2^j \mu(E_{\delta_j}(\rho\zeta)) \\ &\le C_n \sum_{j=n-2} 2^j M_{\delta_j}\mu(\zeta).\end{aligned}$$

Thus

$$M_{\text{rad}} V_1(\zeta) \le C_n \sum_{j=n-2} 2^j M_{\delta_j}\mu(\zeta).$$

Choose α, $0 < \alpha < 1/(n-2)$. Then

$$M_{\text{rad}} V_1(\zeta) \le C_n \sum_{j=(n-2)}^{\infty} 2^{-\alpha j} 2^{(1+\alpha)j} M_{\delta_j}\mu(\zeta),$$

which since $\sum 2^{-\alpha j} < \infty$

$$\le C_n \sup_{j \ge (n-2)} 2^{(1+\alpha)j} M_{\delta_j}\mu(\zeta).$$

For $\lambda > 0$ let

$$E_\lambda = \{\zeta \in \mathbb{S} : (M_{\text{rad}} V_1)(\zeta) > \lambda\}.$$

By the above,

$$E_\lambda \subset \bigcup_{j=(n-2)}^{\infty} \left\{\zeta \in \mathbb{S} : M_{\delta_j}\mu(\zeta) > \frac{\lambda}{C_n} 2^{-(1+\alpha)j}\right\}.$$

Therefore, by Lemma 9.4.5

$$\sigma(E_\lambda) \le \left(\frac{C_n}{\lambda}\int_{\mathbb{B}} (1-|x|^2)^{n-1} d\mu(x)\right)\left(\sum_{j=(n-2)}^{\infty} 2^{(1+\alpha)j}\delta_j^{(n-1)}\right).$$

Since $2^{(1+\alpha)j}\delta_j^{(n-1)} = 2^{-\beta j}$ where

$$\beta = \frac{n-1}{n-2} - (1+\alpha) = \frac{1-\alpha(n-2)}{n-2} > 0,$$

we have

$$\sigma(E_\lambda) \le \frac{C_n}{\lambda}\int_{\mathbb{B}} (1-|x|^2)^{n-1} d\mu(x).$$

□

Proof of Theorem 9.4.1. Let μ be a non-negative regular Borel measure on $\mathbb{B}$ satisfying (9.2.4) and let V_1 be as defined in Lemma 9.4.6. Let $\epsilon > 0$ be arbitrary and let C be the constant of Lemma 9.4.6. Choose R, $0 < R < 1$, such that

$$C \int_{A_R} (1-|y|^2)^{n-1} d\mu(y) < \epsilon^2.$$

Let μ_R be the measure μ restricted to A_R and let

$$V_R(x) = \int_{E(x)} G_h(x,y)\, d\mu_R(y).$$

Suppose $y \in E(x)$. Let c_x and r_x be the Euclidean center and radius of $E(x)$ as given in Theorem 2.2.2. Then

$$|y| \geq |c_x| - r_x = \frac{2|x| - 1}{2 - |x|}.$$

Hence for all x, $1 > |x| > (2R+1)/(R+2)$ we have $y \in A_R$. Hence there exists R', $0 < R' < 1$, such that $E(x) \subset A_R$ for all x, $|x| > R'$. Hence by Lemma 9.4.6,

$$\begin{aligned}\sigma(\{\zeta \in \mathbb{S} : \limsup_{r\to 1} V_1(r\zeta) > \epsilon\}) &\leq \sigma(\{\zeta \in \mathbb{S} : (M_{\text{rad}} V_R)(\zeta) > \epsilon\}) \\ &\leq \frac{C}{\epsilon}\int_{A_R} (1 - |y|^2)^{n-1} d\mu(y) < \epsilon.\end{aligned}$$

Since $\epsilon > 0$ was arbitrary, the result follows. □

Example 9.4.7 We now provide an example of a measure μ satisfying (9.2.4) for which

$$\lim_{\substack{x\to\zeta \\ x\in\Gamma_\alpha(\zeta)}} G_\mu(x) = +\infty$$

for every $\zeta \in \mathbb{S}$ and $\alpha > 1$. Let $\{x_j\}$ be a countable infinite subset of $\mathbb{B}$ with $|x_j| \to 1$ such that $\Gamma_\alpha(\zeta)$ contains infinitely many x_j for every $\zeta \in \mathbb{S}$ and $\alpha > 1$. Such a sequence can be obtained by taking a countable dense subset $\{\zeta_j\}$ of $\mathbb{S}$ and a sequence $\{r_j\}$ increasing to 1, and setting $x_j = r_j\zeta_j$. Now choose $c_j > 0$ such that

$$\sum_{j=1}^{\infty} c_j(1 - |x_j|^2)^{n-1} < \infty.$$

Define the measure μ on $\mathbb{B}$ by

$$\mu = \sum_{j=1}^{\infty} c_j \delta_{x_j},$$

where δ_a is point mass measure at a. Then G_μ is a potential on $\mathbb{B}$ satisfying $G_\mu(x_j) = \infty$ for all j.

9.5 Non-tangential Limits of $\mathcal{H}$-Subharmonic Functions

In Corollary 9.3.6 we proved that if $f \in L^q(\mathbb{B}, \nu)$ for some $q > n$, then

$$\lim_{|x|\to 1} V_f(x) = 0,$$

where

$$V_f(x) = \int_{\mathbb{B}} G_h(x,y)f(y)d\nu(y).$$

If f is a non-negative measurable function on $\mathbb{B}$ satisfying

$$\int_{\mathbb{B}} (1-|x|^2)^{n-1} f(x)d\tau(x) < \infty, \tag{9.5.1}$$

then the **Green potential** G_f of f is the $\mathcal{H}$-superharmonic function on $\mathbb{B}$ defined by

$$G_f(x) = \int_{\mathbb{B}} G_h(x,y)f(y)d\tau(y). \tag{9.5.2}$$

When $n = 2$, M. Arsove and A. Huber [9] provided sufficient conditions for the existence of non-tangential limits of subharmonic functions in the unit disc. The results were subsequently extended by J. Cima and C. S. Stanton to $\mathcal{M}$-subharmonic functions on the unit ball in $\mathbb{C}^m$ [16]. We modify their techniques in proving analogous results for $\mathcal{H}$-subharmonic functions on $\mathbb{B}$.

Theorem 9.5.1 *Let f be a non-negative measurable function on $\mathbb{B}$ satisfying (9.5.1). If in addition*

$$\int_{\mathbb{B}} (1-|x|^2)^{n-1} f^q(y)d\tau(y) < \infty$$

for some $q > \frac{n}{2}$, then

$$\lim_{\substack{x\to\zeta \\ x\in\Gamma_\alpha(\zeta)}} G_f(x) = 0 \quad \textit{for almost every } \zeta \in \mathbb{S}.$$

As a consequence of the Riesz decomposition theorem and Theorem 8.3.3 we also have the following.

Corollary 9.5.2 *Let f be an $\mathcal{H}$-subharmonic function on $\mathbb{B}$ with an $\mathcal{H}$-harmonic majorant on $\mathbb{B}$. If the Riesz measure μ_f of f is absolutely continuous and satisfies*

$$\int_{\mathbb{B}} (1-|x|^2)^{n-1} (\Delta_h f(x))^q d\tau(x) < \infty$$

for some $q > \frac{n}{2}$, then f has non-tangential limits at almost every $\zeta \in \mathbb{S}$.

For a non-negative measure μ on $\mathbb{B}$, $\alpha > 1$, and $\zeta \in \mathbb{S}$, set

$$S_\alpha^*\mu(\zeta) = \mu(\Gamma_\alpha(\zeta)). \tag{9.5.3}$$

Also, for $x \in \mathbb{B}$, set

$$\widetilde{\Gamma}_\alpha(x) = \{\eta \in \mathbb{S} : x \in \Gamma_\alpha(\zeta)\}. \tag{9.5.4}$$

For the proof of Theorem 9.5.4 we require the following lemma.

Lemma 9.5.3 *For $x \in \mathbb{B}$, set $x = |x|\zeta$, $\zeta \in \mathbb{S}$. Then*

(a) *$\widetilde{\Gamma}_\alpha(x) \subset S(\zeta, 2\alpha(1-|x|^2))$ for all $\alpha > 1$.*

(b) *Given $\alpha > 1$, there exists $c > 0$ such that $S(\zeta, c(1-|x|^2)) \subset \widetilde{\Gamma}_\alpha(x)$ for all x, $|x| > \frac{1}{2}$.*

Proof. For (a), if $\eta \in \widetilde{\Gamma}_\alpha(x)$, then

$$|\eta - \zeta| \le |\eta - x| + |x - \zeta| < 2\alpha(1-|x|^2).$$

For the proof of (b), suppose $\eta \in S(\zeta, c(1-|x|^2))$ for $c > 0$ to be determined. Then

$$\begin{aligned} |x-\eta| &\le ||x|\zeta - \zeta| + |\zeta - \eta| \\ &< (1-|x|) + c(1-|x|^2) \\ &= \left(c + \frac{1}{1+|x|}\right)(1-|x|^2), \end{aligned}$$

which for $|x| > \frac{1}{2}$

$$\le (c + \tfrac{2}{3})(1-|x|^2).$$

Now given $\alpha > 1$ choose $c > 0$ such that $c + \frac{2}{3} < \alpha$. □

Theorem 9.5.4 *Let μ be a non-negative regular Borel measure on $\mathbb{B}$. Then*

$$\int_{\mathbb{B}} (1-|x|^2)^{n-1} d\mu(x) < \infty \quad \textit{if and only if} \quad S_\alpha^*\mu \in L^1(\mathbb{S}).$$

If this is the case, then for every $\alpha > 1$, $\mu(\Gamma_\alpha(\zeta)) < \infty$ for almost every $\zeta \in \mathbb{S}$, and thus

$$\lim_{R\to 1} \mu(\Gamma_\alpha(\zeta) \cap A_R) = 0 \quad \textit{for almost every } \zeta \in \mathbb{S}.$$

Proof. By Tonelli's theorem,

$$\begin{aligned} \int_{\mathbb{S}} S_\alpha^*(\zeta) d\sigma(\zeta) &= \int_{\mathbb{S}} \int_{\mathbb{B}} \chi_{\Gamma_\alpha(\zeta)}(x) d\mu(x) d\sigma(\zeta) \\ &= \int_{\mathbb{B}} \int_{\mathbb{S}} \chi_{\widetilde{\Gamma}_\alpha(x)}(\zeta) d\sigma(\zeta) d\mu(x) \\ &= \int_{\mathbb{B}} \sigma(\widetilde{\Gamma}_\alpha(x)) d\mu(x). \end{aligned}$$

By Lemma 9.5.3, $\sigma(\widetilde{\Gamma}_\alpha(x)) \le \sigma(S(\frac{x}{|x|}, 2\alpha(1-|x|^2)) \le C_\alpha(1-|x|^2)^{n-1}$, and thus $S^*_\alpha\mu \in L^1(\mathbb{S})$ whenever μ satisfies (9.2.4).

Conversely, since μ is regular, μ satisfies (9.2.4) if and only if

$$\int_{A_{\frac{1}{2}}} (1-|x|^2)^{n-1} d\mu(x) < \infty.$$

Thus by Lemma 9.5.3,

$$\int_{A_{\frac{1}{2}}} (1-|x|^2)^{n-1} d\mu(x) \le C \int_{A_{\frac{1}{2}}} \sigma(\widetilde{\Gamma}_\alpha(x)) d\mu(x) \le C \int_{\mathbb{S}} S^*_\alpha(\zeta) d\sigma(\zeta).$$

The second part of the theorem follows immediately from the fact that $S^*_\alpha\mu \in L^1(\mathbb{S})$. □

Proof of Theorem 9.5.1. Suppose f is a non-negative measurable function satisfying the hypothesis of Theorem 9.5.1. Set $G_f = V_1 + V_2$, where

$$V_1(x) = \int_{E(x)} G_h(x,y) f(y) d\tau(y) \quad \text{and} \quad V_2(x) = \int_{\mathbb{B}\setminus E(x)} G_h(x,y) f(y) d\tau(y).$$

Since $d\mu = f d\tau$ satisfies (9.2.4), by Theorem 9.4.2 the function $V_2(x)$ has non-tangential limits at almost every $\zeta \in \mathbb{S}$.

Consider the function

$$V_1(x) = \int_{E(x)} g_h(\varphi_x(y)) f(y) d\tau(y),$$

which by the change of variable $w = \varphi_x(y)$ and the invariance of τ

$$= \int_{B_{\frac{1}{2}}} g_h(w) f(\varphi_x(w)) d\tau(w).$$

Let p denote the conjugate exponent of q. Then by Hölder's inequality

$$V_1(x) \le \left[\int_{B_{\frac{1}{2}}} g_h^p(w) d\tau(w)\right]^{1/p} \left[\int_{B_{\frac{1}{2}}} f^q(\varphi_x(w)) d\tau(w)\right]^{1/q}.$$

By inequality (3.2.3) and integration in polar coordinates

$$\int_{B_{\frac{1}{2}}} g_h^p(w) d\tau(w) \le C \int_0^{\frac{1}{2}} r^{n-1-p(n-2)} dr.$$

The above integral, however, is finite if and only if $p < n/(n-2)$, that is, $q > n/2$. Thus for $q > n/2$,

$$V_1(x) \le C\left[\int_{B_{\frac{1}{2}}} f^q(\varphi_x(w))d\tau(w)\right]^{1/q} = C\left[\int_{E(x)} f^q(y)d\tau(y)\right]^{1/q}.$$

Suppose $x \in \Gamma_\alpha(\zeta)$, $\alpha > 1$. If $y \in E(x)$, then $|y| \ge (2|x|-1)/(2-|x|)$. Hence, as in the proof of Theorem 9.4.1, given R, $0 < R < 1$, $E(x) \subset A_R$ for all x, $|x| > (2R+1)/(R+2)$. Let c_x and r_x denote the Euclidean center and radius of $E(x)$. Then for $y \in E(x)$,

$$\begin{aligned}
|y-\zeta| &\le |y-c_x| + |c_x - x| + |x-\zeta| \\
&\le 2r_x + \alpha(1-|x|) \\
&\le \tfrac{4}{3}(1-|x|^2) + \alpha(1-|x|) \le (\alpha + \tfrac{8}{3})(1-|x|).
\end{aligned}$$

Hence, given $0 < R < 1$, $\alpha > 1$, there exists R', $0 < R' < 1$, and $\beta_\alpha > 1$ such that $E(x) \subset \Gamma_{\beta_\alpha}(\zeta) \cap A_R$. As a consequence,

$$V_1(x) \le C\left[\int_{\Gamma_{\beta_\alpha}(\zeta)\cap A_R} f^q(y)d\tau(y)\right]^{1/q}$$

for all $x \in \Gamma_\alpha(\zeta)$, $|x| > R'$. The result now follows by Theorem 9.5.4 with $d\mu = f^q d\tau$. □

Our final theorem of this section, although valid for all $n \ge 2$, is of particular interest when $n = 2$.

Theorem 9.5.5 *Let $\{a_j\}$ be a sequence in $\mathbb{B}$ satisfying*

$$\sum_{j=1}^{\infty}(1-|a_j|^2)^{n-1} < \infty, \tag{9.5.5}$$

and let $\mu = \sum_{j=1}^{\infty}\delta_{a_j}$, where δ_a denotes point mass measure at a. Then for every $\alpha > 1$,

$$\lim_{\substack{x\to\zeta \\ x\in\Gamma_\alpha(\zeta)}} G_\mu(x) = 0 \quad \textit{for almost every } \zeta \in \mathbb{S}.$$

Proof. As a consequence of Theorem 9.4.2, we again only need to consider the function

$$V_1(x) = \int_{E(x)} G_h(x,y)d\mu(y).$$

For each $\alpha > 1$,

$$\mu(\Gamma_\alpha(\zeta)) = |\{j \in \mathbb{N} : a_j \in \Gamma_\alpha(\zeta)\}|,$$

where for each subset J of $\mathbb{N}$, $|J|$ denotes the number of elements in J. If $\mu(\Gamma_\alpha(\zeta)) < \infty$, then $\Gamma_\alpha(\zeta)$ contains only a finite number of a_j. Consequently, for such a ζ,

$$\lim_{\substack{x\to\zeta \\ x\in\Gamma_\alpha(\zeta)}} V_1(x) = 0.$$

But by Theorem 9.5.4, $\mu(\Gamma_\alpha(\zeta)) < \infty$ for almost every $\zeta \in \mathbb{S}$. □

Remark 9.5.6 *In the case when $n = 2$, the sequence $\{a_j\}$ is a subset of a unit disc $\mathbb{D}$, and as such we will assume they are in complex form. Writing $z = x_1 + ix_2$, $(x_1, x_2) \in \mathbb{B}$, we have*

$$G_\mu(z) = \sum_{j=1}^{\infty} \log \left| \frac{1 - \overline{a}_j z}{z - a_j} \right| = -\log |B(z)|,$$

where B is the Blaschke product with (non-zero) zeros $\{a_j\}$ given by

$$B(z) = \prod_{j=1}^{\infty} \frac{|a_j|}{a_j} \frac{z - a_j}{1 - \overline{a}_j z}.$$

When $n = 2$, the hypothesis (9.5.5) is known as the Blaschke condition, which is necessary and sufficient for a bounded analytic function in $\mathbb{D}$ having prescribed zeros $\{a_j\}$.

Example 9.5.7 We conclude this section by providing an example to show that the exponent $q > n/2$ is the best possible. Specifically, we construct an example of a function f satisfying the hypothesis of Theorem 9.5.1 with $q = n/2$ for which

$$\lim_{\substack{x\to\zeta \\ x\in\Gamma_\alpha(\zeta)}} G_f(x) = +\infty \quad \text{for every } \zeta \in \mathbb{S}. \tag{9.5.6}$$

As in the example of the previous section, let $\{x_j\}$ be a countably infinite subset of $\mathbb{B}$ with $|x_j| \to 1$ as $j \to \infty$ such that $\Gamma_\alpha(\zeta)$ contains infinitely many x_j for every $\zeta \in \mathbb{S}$ and $\alpha > 1$.

For each j, choose δ_j, $0 < \delta_j < \frac{1}{2}$, and $c_j > 0$ such that

(a) the family $\{E(x_j, \delta_j)\}$ is pairwise disjoint, and

(b) $\displaystyle\sum_{j=1}^{\infty} (1 - |x_j|^2) c_j^{2/n} < \infty.$

Let $\{a_j\}$ be a sequence of positive numbers with $a_j \to \infty$. For each j, choose a non-negative measurable function f_j satisfying

(1) $\operatorname{supp} f_j \subset E(x_j, \delta_j)$,

(2) $\displaystyle\int f_j^{n/2} d\tau < c_j$, and

(3) $\displaystyle\int_{\mathbb{B}} f_j(y) G_h(x_j, y) d\tau(y) > a_j$.

Clearly for each j we can find an f_j satisfying (1) and (2). Fix a j and let $E_j = E(x_j, \delta_j)$. If we cannot find an f_j satisfying (1)–(3), then we have

$$\int_{E_j} f(y) G_h(x_j, y) d\tau(y) \le a_j c_j^{2/n}$$

for all non-negative measurable functions satisfying $\int_{E_j} f^{n/2} d\tau \le 1$. By duality, this, however, implies that $y \to G_h(x_j, y) \in L^{n/(n-2)}(E_j, \tau)$, which is a contradiction.

Let $f = \sum f_j$. We now show that f satisfies the hypothesis of Theorem 9.5.1 with $q = n/2$. First, by (3.3.8)

$$\tau(E(x_j, \delta_j)) \approx \frac{\delta_j^n}{(1-\delta_j^2)^{n-1}} \le C_n.$$

Also, for $y \in E(x_j, \delta_j)$ we have $(1 - |y|^2) \le 3(1 - |x_j|^2)$. Thus by Hölder's inequality

$$\begin{aligned}\int_{\mathbb{B}} (1-|y|^2)^{n-1} f(y) d\tau(y) &\le 3 \sum_{j=1}^{\infty} (1-|x_j|^2)^{n-1} \int_{E_j} f_j d\tau \\ &\le 3 \sum_{j=1}^{\infty} (1-|x_j|^2)^{n-1} (\tau(E_j))^{(n-2)/n} \left(\int_{E_j} f_j^{n/2} d\tau\right)^{2/n} \\ &\le 3C_n \sum_{j=1}^{\infty} (1-|x_j|^2)^{n-1} c_j^{2/n} < \infty.\end{aligned}$$

Thus the measure $d\mu = f d\tau$ satisfies (9.5.1). Also,

$$\int_{\mathbb{B}} (1-|y|^2) f^{n/2}(y) d\tau(y) \le 3 \sum_{j=1}^{\infty} (1-|x_j|^2)^{n-1} c_j.$$

By convergence of the series in (b), there exists j_o such that $(1-|x_j|^2) c_j^{2/n} \le 1$ for all $j \ge j_o$. Hence, for $j \ge j_o$, and all $n \ge 2$,

$$(1-|x_j|^2)^{n-1} c_j \le \left[(1-|x_j|^2) c_j^{2/n}\right]^{n/2} \le (1-|x_j|^2) c_j^{2/n}.$$

Thus

$$\int_{\mathbb{B}} (1 - |y|^2)^{n-1} f^{n/2}(y) d\tau(y) \le C \sum_{j=1}^{\infty} (1 - |x|_j^2) c_j^{2/n} < \infty.$$

Therefore f satisfies the hypothesis of Theorem 9.5.1 with $q = \frac{n}{2}$. Finally, since

$$G_f(x_j) \ge \int_{\mathbb{B}} f_j(y) G_h(x_j, y) d\tau(y) > a_j,$$

the potential G_f satisfies (9.5.6) at every $\zeta \in \mathbb{S}$. By choosing the f_j to be C^∞ functions we obtain the existence of a C^∞ potential G_f satisfying (9.5.6) at every $\zeta \in \mathbb{S}$.

9.6 Exercises

9.6.1. Construct a measure μ on $\mathbb{B} \subset \mathbb{R}^3$ such that e_1 is a limit point of supp μ, $\int_{\mathbb{B}} (1 - |x|^2)^2 d\mu(x) < \infty$, with $G_\mu(re_1) = +\infty$ for all r, $0 < r < 1$.

9.6.2. (*) In [8] D. H. Armitage provided an example of a Euclidean harmonic function h on $\mathbb{B}$ such that $|h| \notin L^p(\mathbb{B})$ for any $p > 0$. Construct an example of an $\mathcal{H}$-harmonic function h on $\mathbb{B}$ such that $|h| \notin L^p(\mathbb{B}, \nu)$ for any $p > 0$.

Definition 9.6.1 *A function $V(x)$ on $\mathbb{B}$ has* **non-tangential limit** L **at** $\zeta \in \mathbb{S}$ **in** L^p, $p > 0$, *if for every $\alpha > 1$,*

$$\lim_{\delta \to 0} \frac{1}{\nu(\Gamma_{\alpha,\delta}(\zeta))} \int_{\Gamma_{\alpha,\delta}(\zeta)} |V(x) - L|^p d\nu(x) = 0.$$

where $\Gamma_{\alpha,\delta}(\zeta) = \Gamma_\alpha(\zeta) \cap B(\zeta, \delta)$.

In [98] L. Ziomek proved the following theorem. *Suppose that V is subharmonic in $\mathbb{B}$ and*

$$\sup_{0<r<1} \int_{\mathbb{S}} V^+(rt) d\sigma(t) < \infty. \tag{9.6.1}$$

Then V has a non-tangential limit in L^p, $1 \le p < n/(n-2)$ at almost every $\zeta \in \mathbb{S}$. This limit coincides with the ordinary radial limit of V. An analogue of this result for $\mathcal{M}$-harmonic functions on the unit ball in $\mathbb{C}^m$ has also been proved by J. Cima and C. S. Stanton [16].

9.6.3. (*) Prove the following **Conjecture.** Let V be $\mathcal{H}$-subharmonic in $\mathbb{B}$ satisfying (9.6.1). Then V has a non-tangential limit in L^p, $0 < p < n/(n-2)$, at almost every $\zeta \in \mathbb{S}$. This limit coincides with the ordinary radial limit of V.

9.6.4. **Tangential limits of potentials.** For $\alpha > 1$, $\tau > 1$, $\zeta \in \mathbb{S}$, set

$$\Gamma_{\alpha,\tau}(\zeta) = \{x \in \mathbb{B} : |x - \zeta|^{\tau} < \alpha(1 - |x|)\}.$$

The set $\Gamma_{\alpha,\tau}(\zeta)$ is called a **tangential approach region** at $\zeta \in \mathbb{S}$. It can be shown that for $\tau > 1$ the regions $\Gamma_{\alpha,\tau}(\zeta)$ have tangential contact in all directions at $\zeta \in \mathbb{S}$ (see [86]).

Definition 9.6.2 *A function V on $\mathbb{B}$ is said to have* **tangential limit** L **of order** τ *at* $\zeta \in \mathbb{S}$ *if*

$$\lim_{\substack{x \to \zeta \\ x \in \Gamma_{\alpha,\tau}(\zeta)}} V(x) = L$$

for every $\alpha > 1$. Likewise, V has ***tangential limit** L **of order** τ **in** L^p* *at $\zeta \in \mathbb{S}$ if*

$$\lim_{\delta \to 0} \frac{1}{\nu(\Gamma_{\alpha,\tau,\delta}(\zeta))} \int_{\Gamma_{\alpha,\tau,\delta}(\zeta)} |V(x) - L|^p d\nu(x) = 0,$$

where $\Gamma_{\alpha,\tau,\delta}(\zeta) = \Gamma_{\alpha,\tau}(\zeta) \cap B(\zeta, \delta)$.

Tangential boundary limits of harmonic functions or Green potentials have been considered by many authors, including H. Aikawa [6], M. Arsove and H. Huber [9], Y. Mizuta [59, 60], A. Nagel, W. Rudin, and J. H. Shapiro [61], J.-M. G. Wu [95], and the author [83, 86], among many others.

Investigate the following two conjectures for $\mathcal{H}$-superharmonic functions on $\mathbb{B}$.

(a) **Conjecture 1.** If f is a non-negative measurable function on $\mathbb{B}$ satisfying

$$\int_{\mathbb{B}} (1 - |x|^2)^{\gamma} f^q(x) d\tau(x) < \infty$$

for some γ, $0 < \gamma < (n - 1)$, and some $q > \frac{n}{2}$, then G_f has tangential limit 0 of order τ, $1 \le \tau \le (n-1)/\gamma$, at a.e. $\zeta \in \mathbb{S}$.

(b) **Conjecture 2.** If μ is a non-negative regular Borel measure on $\mathbb{B}$ satisfying

$$\int_{\mathbb{B}} (1 - |x|^2)^{\gamma} d\mu(x) < \infty$$

for some γ, $0 < \gamma < (n - 1)$, then G_μ has tangential limit 0 of order τ, $1 \le \tau \le (n-1)/\gamma$, in L^p, $0 < p < n/(n-2)$, at a.e. $\zeta \in \mathbb{S}$.

9.6.5. (*) **Weighted boundary limits of non-negative $\mathcal{H}$-subharmonic functions.** In [28] F. W. Gehring proved that if w is a non-negative subharmonic function in the unit disc $\mathbb{D}$ satisfying

$$\iint_{\mathbb{D}} w^p(z)dxdy < \infty$$

for some $p > 1$, then

$$\lim_{\substack{z\to\zeta \\ z\in\Gamma_\alpha(e^{i\theta})}} (1-|z|)w^p(z) = 0 \quad \text{for a.e. } \theta.$$

This result was extended by D. Hallenbeck [33] to $0 < p \le 1$, and independently by the author for all $p > 0$, as well as subharmonic functions on bounded domains with C^1 boundary [86]. The paper [86] also contains results concerning tangential approach regions. Analogous results for $\mathcal{M}$-subharmonic functions have also been proved by the author in [85].

Investigate weighted boundary limits of non-negative $\mathcal{H}$-subharmonic functions f on $\mathbb{B}$ satisfying $f \in L^p_\gamma(\tau)$ for some $p > 0$ and some $\gamma > (n-1)\min\{1,p\}$.

9.6.6. (*) **Weighted boundary limits of potentials.** This area of investigation is motivated by an old result of M. Heins concerning weighted limits of $\log|B(z)|$, where $B(z)$ is a convergent Blaschke product in the unit disc $\mathbb{D}$. In [37] M. Heins proved that if B is a Blaschke product in $\mathbb{D}$, then

$$\liminf_{r\to 1}(1-r)\log\frac{1}{|B(re^{i\theta})|} = 0 \tag{9.6.2}$$

for all θ, $0 \le \theta < 2\pi$. This result was extended by the author [80] to potentials on $\mathbb{D}$ in the following theorem.

Theorem 9.6.3 *If G_μ is the potential of a measure μ satisfying (9.2.4) with $n = 2$, then for all curves $\gamma : [0,1) \to \mathbb{D}$ with $\lim_{t\to 1}\gamma(t) = 1$,*

$$\liminf_{t\to 1}(1-|\gamma(t)|)G_\mu(\gamma(t)e^{i\theta}) = 0$$

for all θ, $0 \le \theta < 2\pi$.

Exercise 9.6.1 shows that the analogue of (9.6.2) is false for $n \ge 3$. There are, however, several extensions of Theorem 9.6.3 that are worthy of consideration. The first involves the following question. Suppose F

is a relatively closed subset of $\mathbb{B}$ such that $t \in \mathbb{S}$ is a limit point of F. What are necessary and sufficient conditions on F such that

$$\liminf_{\substack{x \to t \\ x \in F}} (1 - |x|^2)^{n-1} G_\mu(x) = 0? \tag{9.6.3}$$

When $n = 2$, this question was answered by D. H. Luecking [52], and subsequently extended to invariant potentials on the unit ball in $\mathbb{C}^m$ by K. T. Hahn and D. Singman [32]. It was proved that (9.6.3) holds if and only if the capacity of the sets

$$F \cap \{x \in \mathbb{B} : |z - t| < \epsilon\}, \quad \epsilon > 0,$$

is bounded away from zero.

Here, the **capacity** $c(K)$ of a compact set K is defined as

$$c(K) = \sup\{\mu(K) : \operatorname{supp} \mu \subset K \text{ and } G_\mu \leq 1 \text{ on } K\}.$$

For an arbitrary set A,

$$c(A) = \sup\{c(K) : K \text{ is a compact subset of } A\}.$$

In a different direction, one can show that when $n = 2$, Theorem 9.6.3 is equivalent to

$$\liminf_{r \to 1} (1 - r) M_\infty(G_\mu, r),$$

where $M_\infty(G_\mu, r) = \sup_{|z|=r} G_\mu(z)$. Again, Exercise 9.6.1 shows that this is false for $n \geq 3$. The above result was extended by S. Gardiner [25] to Euclidean potentials on $\mathbb{B} \subset \mathbb{R}^n$ (see Exercise 4.8.12), and by the author to invariant potentials [81], [82] on $\mathbb{B} \subset \mathbb{C}^m$.

10

Bergman and Dirichlet Spaces of $\mathcal{H}$-Harmonic Functions

In this chapter we consider Bergman and Dirichlet spaces of $\mathcal{H}$-harmonic functions on $\mathbb{B}$. For $p > 0$ and $\gamma \in \mathbb{R}$, let

$$L^p_\gamma(\tau) = \left\{ f \text{ measurable } : \int_{\mathbb{B}} (1 - |x|^2)^\gamma |f(x)|^p d\tau(x) < \infty \right\},$$

with

$$\|f\|_{\gamma,p} = \left(\int_{\mathbb{B}} (1 - |x|^2)^\gamma |f(x)|^p d\tau(x) \right)^{1/p}. \tag{10.0.1}$$

The $\mathcal{H}$-harmonic **weighted Bergman space** $\mathcal{B}^p_\gamma$, $0 < p < \infty, \gamma \in \mathbb{R}$, is defined as the space of $\mathcal{H}$-harmonic functions f on $\mathbb{B}$ for which $f \in L^p_\gamma(\tau)$. For $\gamma = n$ we obtain the classical Bergman space $\mathcal{B}^p = \mathcal{B}^p_n$ of $\mathcal{H}$-harmonic functions f on $\mathbb{B}$ with $f \in L^p(\nu)$.

Also, for $0 < p < \infty, \gamma \in \mathbb{R}$, $\mathcal{D}^p_\gamma$ denotes the **weighted Dirichlet space** of $\mathcal{H}$-harmonic functions f for which $|\nabla^h f| \in L^p_\gamma$ with

$$\|f\|_{\mathcal{D}^p_\gamma} = |f(0)| + \left(\int_{\mathbb{B}} (1 - |x|^2)^\gamma |\nabla^h f(x)|^p d\tau(x) \right)^{1/p}. \tag{10.0.2}$$

Since $|\nabla^h f(x)| = (1 - |x|^2)|\nabla f(x)$,

$$\mathcal{D}^p_\gamma = \left\{ f \ \mathcal{H}\text{-harmonic } : \int_{\mathbb{B}} (1 - |x|^2)^{\gamma+p-n} |\nabla f(x)|^p d\nu(x) < \infty \right\}.$$

In Section 10.1 we investigate basic properties of the spaces $\mathcal{B}^p_\gamma$ and $\mathcal{D}^p_\gamma$, and in Section 10.2 we provide a brief discussion of Möbius invariant spaces on $\mathbb{B}$. This leads to several open problems concerning the Möbius invariant Hilbert space of $\mathcal{H}$-harmonic functions on $\mathbb{B}$. In Section 10.3 we prove that for $\gamma > n-1$ the spaces $\mathcal{B}^p_\gamma$ and $\mathcal{D}^p_\gamma$ are equivalent for all p, $0 < p < \infty$. Sections 10.4 and 10.5 deal with questions concerning the integrability of functions in $\mathcal{B}^p_\gamma$ and $\mathcal{D}^p_\gamma$ as well as of eigenfunctions of Δ_h. Here we are primarily concerned with the question of given p, $0 < p < \infty$, for what values of γ are the spaces $\mathcal{B}^p_\gamma$ and $\mathcal{D}^p_\gamma$ non-trivial? Finally, in Section 10.6 we prove generalizations of three theorems of Hardy and Littlewood for $\mathcal{H}$-harmonic functions on $\mathbb{B}$, and in Section 10.7 we prove a generalization of the Littlewood–Paley inequalities. Other generalizations and results are provided in the exercises.

10.1 Properties of $\mathcal{D}^p_\gamma$ and $\mathcal{B}^p_\gamma$

We first prove that if $f \in \mathcal{B}^p_\gamma$ (or $\mathcal{D}^p_\gamma$), then $f \circ \varphi_a \in \mathcal{B}^p_\gamma$ ($\mathcal{D}^p_\gamma$ respectively). We restrict ourself to the case $\gamma > 0$, which as we will later see will always be necessary in order that these spaces be non-trivial.

Theorem 10.1.1 *Let $f \in \mathcal{B}^p_\gamma$ (respectively $\mathcal{D}^p_\gamma$) with $\gamma > 0$, $0 < p < \infty$, then $f \circ \varphi_a \in \mathcal{B}^p_\gamma$ (respectively $\mathcal{D}^p_\gamma$) with*

$$\|f \circ \varphi_a\|_{\gamma,p} \le \left(\frac{1+|a|}{1-|a|}\right)^{\frac{\gamma}{p}} \|f\|_{\gamma,p}, \tag{10.1.1}$$

or

$$\|f \circ \varphi_a\|_{\mathcal{D}^p_\gamma} \le C \left(\frac{1+|a|}{1-|a|}\right)^{\frac{\gamma}{p}+1} \|f\|_{\mathcal{D}^p_\gamma}, \tag{10.1.2}$$

for some constant C independent of f and a.

Proof. For $a \in \mathbb{B}$,

$$\|f \circ \varphi_a\|^p_{\gamma,p} = \int_{\mathbb{B}} (1-|x|^2)^\gamma |f(\varphi_a(x))|^p d\tau(x),$$

which by the invariance of τ

$$= \int_{\mathbb{B}} (1-|\varphi_a(x)|^2)^\gamma |f(x)|^p d\tau(x).$$

However, by (2.1.7),

$$\begin{aligned}(1-|\varphi_a(x)|^2)^\gamma &= \frac{(1-|x|^2)^\gamma(1-|a|^2)^\gamma}{\rho(x,a)^\gamma} \\ &\le \frac{(1-|x|^2)^\gamma(1-|a|^2)^\gamma}{(1-|x||a|)^{2\gamma}} \\ &\le \frac{(1-|x|^2)^\gamma(1+|a|)^\gamma}{(1-|a|)^\gamma},\end{aligned}$$

from which (10.1.1) follows.

For inequality (10.1.2), as above, by the invariance of ∇^h and τ we have

$$\|f \circ \varphi_a\|_{\mathcal{D}_\gamma^p} \le |f(a)| + \left(\frac{1+|a|}{1-|a|}\right)^{\gamma/p} \|\nabla^h f\|_{\gamma,p}. \tag{10.1.3}$$

By the fundamental theorem of calculus,

$$\begin{aligned} |f(a)| &\le |f(0)| + \int_0^{|a|} \frac{|\nabla^h f(rt)|}{(1-r^2)} dr \\ &\le |f(0)| + \frac{1}{(1-|a|^2)} \sup\{|\nabla^h f(rt)| : 0 < r \le |a|\}. \end{aligned}$$

But for $x \in B_{|a|}$ and $0 < \delta < \frac{1}{2}$ fixed, by Theorem 4.7.4,

$$|\nabla^h f(x)|^p \le C_\delta \int_{E_\delta(x)} |\nabla^h f(y)|^p d\tau(y),$$

which since $(1-|x|^2) \approx (1-|y|^2)$ for $y \in E_\delta(x)$

$$\begin{aligned} &\le \frac{C_\delta}{(1-|x|^2)^\gamma} \int_{E_\delta(x)} (1-|y|^2)^\gamma |\nabla^h f(y)|^p d\tau(y) \\ &\le \frac{C_\delta}{(1-|a|^2)^\gamma} \int_{\mathbb{B}} (1-|y|^2)^\gamma |\nabla^h f(y)|^p d\tau(y). \end{aligned}$$

Therefore,

$$|f(a) \le |f(0)| + \frac{C_\delta}{(1-|a|^2)^{\gamma/p+1}} \|\nabla^h f\|_{\gamma,p},$$

from which (10.1.2) follows. □

In the above we have proved that if $f \in \mathcal{D}_\gamma^p, 0 < p < \infty$, $\gamma > 0$, then for all $a \in \mathbb{B}$,

$$|f(a)| \le \frac{C}{(1-|a|^2)^{\gamma/p+1}} \|f\|_{\mathcal{D}_\gamma^p}, \tag{10.1.4}$$

where C is a constant depending on a fixed δ, $0 < \delta < \frac{1}{2}$. Likewise, if $f \in \mathcal{B}_\gamma^p, 0 < p < \infty$, $\gamma > 0$, then

$$|f(a)| \le \frac{C}{(1-|a|^2)^{\gamma/p}} \|f\|_{\gamma,p}. \tag{10.1.5}$$

It is easy to show that $\|\cdot\|_{\gamma,p}$ (respectively $\|\cdot\|_{\mathcal{D}_\gamma^p}$) is a norm on $\mathcal{B}_\gamma^p$ (respectively $\mathcal{D}_\gamma^p$) when $1 \le p < \infty$, whereas $\|\cdot\|_{\gamma,p}^p$ and $\|\cdot\|_{\mathcal{D}_\gamma^p}^p$ are p-norms when $0 < p < 1$.

As in (7.2.1) define $d_{\mathcal{B}^p_\gamma}$ on $\mathcal{B}^p_\gamma$ by

$$d_{\mathcal{B}^p_\gamma}(f,g) = \begin{cases} \|f-g\|_{\gamma,p}, & 1 \le p < \infty, \\ \|f-g\|^p_{\gamma,p}, & 0 < p < 1. \end{cases} \tag{10.1.6}$$

Then $d_{\mathcal{B}^p_\gamma}$ is a metric on $\mathcal{B}^p_\gamma$, $0 < p < \infty$. The metric $d_{\mathcal{D}^p_\gamma}$ on $\mathcal{D}^p_\gamma$ is defined similarly. As in Section 7.2 we have the following theorem, the proof of which is similar to the proof of Theorem 7.2.2 and thus is omitted.

Theorem 10.1.2 *The spaces $(\mathcal{B}^p_\gamma, d_{\mathcal{B}^p_\gamma})$ and $(\mathcal{D}^p_\gamma, d_{\mathcal{D}^p_\gamma})$ are complete metric spaces for all p, $0 < p < \infty$, and Banach spaces for $1 \le p < \infty$.*

We close this section with two integral formulas for functions in $\mathcal{B}^1_\gamma$, $\gamma > (n-1)$, and $\mathcal{B}^1_n$ respectively. Our first result is an analogue of [79, Theorem 1].

Theorem 10.1.3 *For $f \in \mathcal{B}^1_\lambda$, $\lambda > (n-1)$,*

$$f(a) = \frac{1}{c_\lambda}\int_{\mathbb{B}} (1-|\varphi_a(y)|^2)^\lambda f(y)\,d\tau(y), \quad a \in \mathbb{B},$$

where c_λ is a constant depending only on λ.

Proof. Define the function F on $O(n)$ by

$$F(A) = \int_{\mathbb{B}} (1-|x|^2)^\lambda f(\varphi_a(Ax))\,d\tau(x),$$

which by the change of variables $y = Ax$

$$\begin{aligned} &= \int_{\mathbb{B}} (1-|y|^2)^\lambda f(\varphi_a(y))\,d\tau(y) \\ &= F(I). \end{aligned}$$

Therefore $F(A) = F(I)$ for all $A \in O(n)$. Hence by Fubini's theorem,

$$F(I) = \int_{O(n)} F(A)\,dA = \int_{\mathbb{B}}\int_{O(n)} f(\varphi_a(Ax))\,dA\,d\tau(x),$$

which since f is $\mathcal{H}$-harmonic on $\mathbb{B}$

$$\begin{aligned} &= f(a)\int_{\mathbb{B}} (1-|x|^2)^\lambda d\tau(x) \\ &= c_\lambda f(a), \end{aligned}$$

where $c_\lambda = \int_{\mathbb{B}} (1-|x|^2)^\lambda d\tau(x)$. Therefore,

$$f(a) = \frac{1}{c_\lambda}F(I) = \frac{1}{c_\lambda}\int_{\mathbb{B}} (1-|x|^2)^\lambda f(\varphi_a(x))\,d\tau(x).$$

The conclusion of the theorem now follows by the change of variable $y = \varphi_a(x)$. □

Our second theorem is an analogue of [79, Theorem 2].

Theorem 10.1.4 *If* $f \in \mathcal{B}_n^1$, *then*

$$f(x) = \int_{\mathbb{B}} \frac{(1-|x|^2)^n}{\rho(y,x)^n} f(y)d\nu(y). \tag{10.1.7}$$

Proof. Since f is $\mathcal{H}$-harmonic, by Equation (4.3.3),

$$f(a) = f(\varphi_a(0)) = \int_{\mathbb{B}} f(\varphi_a(x))d\nu(x),$$

which by the change of variable $y = \varphi_a(x)$ and Theorem 3.3.1

$$\begin{aligned} &= \int_{\mathbb{B}} f(y)|J_{\varphi_a}(y)|d\nu(y) \\ &= \int_{\mathbb{B}} \frac{(1-|\varphi_a(y)|^2)^n}{(1-|y|^2)^n} f(y)d\nu(y) \\ &= \int_{\mathbb{B}} \frac{(1-|a|^2)^n}{\rho(y,a)^n} f(y)d\nu(y). \end{aligned}$$

□

When $n = 2$, Equation (10.1.7) becomes

$$f(z) = \int_{\mathbb{D}} \frac{(1-|z|^2)^2}{|1-z\overline{w}|^4} f(w)dA(w). \tag{10.1.8}$$

Definition 10.1.5 *For* $f \in L^1(\mathbb{B}, \nu)$, *define*

$$\mathrm{B}f(x) = \int_{\mathbb{B}} \frac{(1-|x|^2)^n}{\rho(x,y)^n} f(y)d\nu(y), \quad x \in \mathbb{B}.$$

In analogy to the case $n = 2$, the operator **B** will be called the **Berezin transform**. For properties of the Berezin tansform when $n = 2$ the reader is referred to the text [36] by H. Hedenmalm, B. Korenblum, and K. Zhu.

By Theorem 10.1.3, for $\alpha > -1$,

$$f(0) = \frac{1}{c_\alpha} \int_{\mathbb{B}} (1-|y|^2)^\alpha f(y)d\nu(y),$$

where

$$c_\alpha = \int_{\mathbb{B}} (1-|y|^2)^\alpha d\nu(y) = n \int_0^1 r^{n-1}(1-r^2)^\alpha dr = \frac{1}{2} \frac{\Gamma(\frac{n}{2})\Gamma(\alpha+1)}{\Gamma(\frac{n}{2}+\alpha+1)}.$$

Hence as above,

$$\begin{aligned} f(a) = f(\varphi_a(0)) &= \frac{1}{c_\alpha}\int_{\mathbb{B}}(1-|y|^2)^\alpha f(\varphi_a(y))d\nu(y) \\ &= \frac{1}{c_\alpha}\int_{\mathbb{B}}(1-|\varphi_a(x)|^2)^\alpha |J_{\varphi_a}(x)| f(x)d\nu(x) \\ &= \frac{1}{c_\alpha}\int \frac{(1-|x|^2)^\alpha(1-|a|^2)^{\alpha+n}}{\rho(x,a)^{\alpha+n}} f(x)d\nu(x). \end{aligned}$$

Thus, for $f \in L^1_\alpha(\mathbb{B},\nu)$, $\alpha > -1$, we write

$$\mathrm{B}_\alpha f(x) = \frac{1}{c_\alpha}\int_{\mathbb{B}} \frac{(1-|x|^2)^{\alpha+n}(1-|y|^2)^\alpha}{\rho(x,y)^{\alpha+n}} f(y)d\nu(y). \tag{10.1.9}$$

If f is $\mathcal{H}$-harmonic on $\mathbb{B}$ with $f \in L^1_\alpha(\mathbb{B},\nu)$, then $\mathrm{B}_\alpha f = f$. The question posed in Chapter 4 following Remark 4.3.6 can now be restated as follows:

Question. If $f \in L^1(\mathbb{B},\nu)$ and $\mathbf{B}f(x) = f(x)$ for all $x \in \mathbb{B}$, is f $\mathcal{H}$-harmonic on $\mathbb{B}$?

10.2 Möbius Invariant Spaces

In this section we provide a brief introduction to Möbius invariant function spaces on $\mathbb{B}$.

Definition 10.2.1 *Let X be a linear space of functions on $\mathbb{B}$ with a complete* **seminorm** $\|\cdot\|$ *on X, that is, $\|f+g\| \le \|f\| + \|g\|$ and $\|\alpha f\| = |\alpha|\|f\|$ for all $\alpha \in \mathbb{R}$ (or $\mathbb{C}$) and $f, g \in X$. The space $(X, \|\ \|)$ is said to be* **Möbius invariant** *iff $f \circ \phi \in X$ and*

$$\|f\circ\varphi\| = \|f\| \quad \text{for all } f \in X,\ \varphi \in \mathcal{M}.$$

The natural candidates for Möbius invariant spaces of $\mathcal{H}$-harmonic functions on $\mathbb{B}$ are $\mathcal{B}^p_\gamma$ and $\mathcal{D}^p_\gamma$ with $\gamma = 0$. In this case

$$\|f\circ\varphi\|^p_{0,p} = \int_{\mathbb{B}} |f(\varphi(x))|^p d\tau(x),$$

which by the invariance of τ

$$= \int_{\mathbb{B}} |f|^p d\tau = \|f\|^p_{0,p}.$$

Furthermore, for $p \geq 1$ $\|f\|_{0,p}$ is a norm on $\mathcal{B}_0^p$. Likewise, by the invariance of ∇^h and τ,

$$\int_{\mathbb{B}} |\nabla^h f(\varphi(x))|^p d\tau(x) = \int_{\mathbb{B}} |\nabla^h f|^p d\tau.$$

In this case,

$$\||f|\|_{\mathcal{D}_\gamma^p} = \left(\int_{\mathbb{B}} (1 - |x|^2)^\gamma |\nabla^h f(x)|^p d\tau(x)\right)^{1/p}$$

is a seminorm on $\mathcal{D}_\gamma^p$. There is only one problem with these spaces. As we will prove in Section 10.4, for $p \geq 1$,

$$\begin{aligned} &\mathcal{B}_\gamma^p \neq \{0\} && \text{if and only if } \gamma > n-1, \text{ and} \\ &\mathcal{D}_\gamma^p \neq \{\text{constants}\} && \text{if and only if } \gamma > (n-1) - p. \end{aligned}$$

Thus $\mathcal{B}_0^p = \{0\}$ for all $p \geq 1$, and thus none of the spaces $\mathcal{B}_0^p$ is a Möbius invariant Banach space.

The spaces $\mathcal{D}_0^p$, however, are all non-trivial Möbius invariant Banach spaces whenever $p > (n-1)$. In particular, when $n = 2$, $\mathcal{D}_0^p$ is a Möbius invariant Banach space for all $p > 1$. Furthermore, when $p = 2$,

$$\mathcal{D}_0^2 = \left\{h \text{ harmonic on } \mathbb{D} : \int_{\mathbb{D}} |\nabla h(z)|^2 dA(z) < \infty\right\}$$

is a Möbius invariant Hilbert space on $\mathbb{D}$. Note, if $h = \operatorname{Re} f$ where f is analytic on $\mathbb{D}$, then since $|\nabla h| = \sqrt{2}|f'(z)|$, the Dirichlet space of harmonic and analytic functions is equivalent. Furthermore, J. Arazy and S. Fisher [7] proved that when $n = 2$ (real dimension), the Dirichlet space $\mathcal{D}_0^2$ was unique among Möbius invariant Hilbert spaces of analytic functions on $\mathbb{D}$. However, for $n \geq 3$, the spaces $\mathcal{D}_0^2$ are trivial. Similar results are also true for the analogous spaces of holomorphic functions on the unit ball in $\mathbb{C}^m$, $m \geq 1$.

K. Zhu [97] and independently M. Peloso [67] proved that for $m \geq 2$, there exists a unique Möbius invariant Hilbert space H of holomorphic functions on the unit ball in $\mathbb{C}^m$. Furthermore, Zhu provided the following characterization of the space H and the Möbius invariant inner product on H. Zhu proved that

$$H = \left\{f(z) = \sum_\alpha a_\alpha z^\alpha : \sum_\alpha |a_\alpha|^2 \frac{\alpha!}{|\alpha|!}|\alpha| < \infty\right\}$$

with Möbius invariant (semi-) inner product

$$\langle f, g\rangle = c \sum_\alpha a_\alpha \overline{b_\alpha} \frac{\alpha!}{|\alpha|!}|\alpha|.$$

In the above, for a multi-index $\alpha = (\alpha_1, \ldots, \alpha_m)$, $\alpha! = \alpha_1! \cdots \alpha_m!$ and $|\alpha| = \alpha_1 + \cdots + \alpha_m$. Also, Peloso proved that the dual of H is the Bloch space $\mathcal{B}$ consisting of the space of holomorphic functions f on $\mathbb{B}$ for which $\sup |\widetilde{\nabla} f(z)| < \infty$, where $\widetilde{\nabla}$ is the invariant gradient on $\mathcal{B}$. Furthermore, Peloso also proved that the reproducing kernel[1] of H is given by

$$K(z, w) = \log \frac{1}{(1 - \langle z, w\rangle)}$$

where $\langle z, w\rangle$ is the usual inner product on $\mathbb{C}^m$.

This leads to three interesting questions.

(1) For $n \geq 3$, does there exist a Möbius invariant Hilbert space H of $\mathcal{H}$-harmonic functions on $\mathbb{B}$?

(2) What is the characterization of H?

(3) What is the reproducing kernel of H?

10.3 Equivalence of $\mathcal{B}_\gamma^p$ and $\mathcal{D}_\gamma^p$ for $\gamma > (n-1)$

Our main result of this section is Theorem 10.3.3 in which we prove that the spaces $\mathcal{B}_\gamma^p$ and $\mathcal{D}_\gamma^p$ are equivalent for all p, $0 < p < \infty$, and all $\gamma > n - 1$. We first prove the following lemma.

Lemma 10.3.1 *For f measurable on $\mathbb{B}$, $\gamma \in \mathbb{R}$, and $0 < \delta < \frac{1}{2}$,*

$$\int_{\mathbb{B}} (1 - |x|^2)^\gamma |f(x)|\, d\tau(x) \approx \int_{\mathbb{B}} (1 - |w|^2)^\gamma \left[\int_{E_\delta(w)} |f(x)|\, d\tau(x) \right] d\tau(w).$$

Proof. For $E \subset \Omega$, let χ_E denote the characteristic function of E. Thus

$$\int_{\mathbb{B}} (1 - |y|^2)^\gamma \left[\int_{E_\delta(y)} |f(x)|\, d\tau(x) \right] d\tau(y)$$
$$= \int_{\mathbb{B}} \int_{\mathbb{B}} (1 - |y|^2)^\gamma \chi_{E_\delta(y)}(x) |f(x)|\, d\tau(x)\, d\tau(y),$$

which by Fubini's theorem

$$= \int_{\mathbb{B}} |f(x)| \left[\int_{\mathbb{B}} \chi_{E_\delta(y)}(x) (1 - |y|^2)^\gamma d\tau(y) \right] d\tau(x).$$

Since $|\varphi_y(x)| = |\varphi_x(y)|$, we have $\chi_{E_\delta(y)}(x) = \chi_{E_\delta(x)}(y)$. Also, for $y \in E_\delta(x)$, $(1 - |y|^2)^\gamma \approx (1 - |x|^2)^\gamma$. Therefore,

[1] See Exercise 10.8.11 for the definition of reproducing kernel.

$$\int_{\mathbb{B}} \chi_{E_\delta(y)}(x)(1-|y|^2)^\gamma d\tau(y) \approx (1-|x|^2)^\gamma \int_{E_\delta(x)} d\tau(y) = (1-|x|^2)^\gamma \tau(B_\delta),$$

from which the result now follows. □

Lemma 10.3.2 *Let g be a non-negative locally integrable function on* $\mathbb{B}$.

(a) *If g is quasi-nearly* $\mathcal{H}$*-subharmonic, then g is quasi-nearly subharmonic, that is, for all* $a \in \mathbb{B}$,

$$g(a) \le \frac{C}{r^n} \int_{B(a,r)} g(x) d\nu(x) \tag{10.3.1}$$

for all $r < r_o$ *with* $\overline{B(a,r)} \subset \mathbb{B}$ *and some constant C.*

(b) *If g is quasi-nearly subharmonic, then for all* ρ, $0 < \rho < 1$, g_ρ *is quasi-nearly subharmonic with constant C independent of* ρ.

Proof. (a) Since g is quasi-nearly $\mathcal{H}$-subharmonic, for all δ, $0 < \delta < \frac{1}{2}$, there exists a constant C independent of g and δ, such that

$$g(a) \le \frac{C}{\delta^n} \int_{E(a,\delta)} g(x) \frac{d\nu(x)}{(1-|x|^2)^n}.$$

Since $\delta < \frac{1}{2}$, by Exercise 2.4.1,

$$\frac{1}{3}(1-|a|^2) \le (1-|x|^2) \le 3(1-|a|^2)$$

for all $x \in E(a,\delta)$. Therefore

$$g(a) \le \frac{2^n C}{(2\delta(1-|a|^2))^n} \int_{E(a,\delta)} g(x) d\nu(x).$$

But by Exercise 4.8.3,

$$E(a,\delta) \subset B(a, c_\delta(1-|a|^2))$$

with $c_\delta = \delta/(1-\delta)$. Since $\delta < \frac{1}{2}$, $c_\delta < 2\delta$. Let r, $0 < r < (1-|a|)$. Choose $\delta < \frac{1}{2}$ such that $2\delta(1-|a|^2) = r$. Then

$$g(a) \le \frac{2^n C}{r^n} \int_{B(a,r)} g(x) d\nu(x),$$

which proves (10.3.1).

(b) Suppose g is quasi-nearly subharmonic. Fix $\rho, 0 < \rho < 1$. Then

$$g_\rho(a) \le \frac{C}{r^n} \int_{B(\rho a,r)} g(x) d\nu(x),$$

which by the change of variable $x = \rho y$

$$= C\frac{\rho^n}{r^n}\int_{B(a,\frac{r}{\rho})} g_\rho(y)dv(y)$$
$$= \frac{C}{\delta^n}\int_{B(a,\delta)} g_\rho(y)dv(y).$$

□

Theorem 10.3.3 *Let f be $\mathcal{H}$-harmonic on $\mathbb{B}$.*

(a) *For all p, $0 < p < \infty$, and $\gamma \in \mathbb{R}$, there exists a constant C, independent of f, such that*

$$\int_{\mathbb{B}} (1-|x|^2)^\gamma |\nabla^h f(x)|^p d\tau(x) \le C\int_{\mathbb{B}} (1-|x|^2)^\gamma |f(x)|^p d\tau(x).$$

(b) *For $\gamma > (n-1)$ and $0 < p < \infty$, there exists a constant C, independent of f, such that*

$$\int_{\mathbb{B}} (1-|x|^2)^\gamma |f(x)|^p d\tau(x) \le C\left[|f(0)|^p + \int_{\mathbb{B}} (1-|x|^2)^\gamma |\nabla^h f(x)|^p d\tau(x)\right].$$

Proof. (a) Fix δ, $0 < \delta < \frac{1}{2}$. Then by Theorem 4.7.4(b)

$$|\nabla^h f(x)|^p \le C_{\delta,p}\int_{E_\delta(x)} |f(y)|^p d\tau(y).$$

Therefore,

$$\int_{\mathbb{B}} (1-|x|^2)^\gamma |\nabla^h f(x)|^p d\tau(x) \le C\int_{\mathbb{B}} (1-|x|^2)^\gamma \left[\int_{E_\delta(x)} |f(y)|^p d\tau(y)\right] d\tau(x),$$

which by Lemma 10.3.1

$$\le C\int_{\mathbb{B}} (1-|x|^2)^\gamma |f(x)|^p d\tau(x).$$

This proves (a).

(b) **(i) The case** $1 < p < \infty$**.** By the fundamental theorem of calculus,

$$|f(x) - f(0)| \le \int_0^{|x|} |\nabla f(t\zeta)|dt \le \int_0^{|x|} \frac{|\nabla^h f(t\zeta)|}{(1-t^2)}dt, \tag{10.3.2}$$

where $x = |x|\zeta$, $\zeta \in \mathbb{S}$. Hence for $|x| \le \frac{1}{2}$,

$$|f(x) - f(0)| \le C\int_0^{|x|} |\nabla^h f(t\zeta)|dt \le C\sup\{|\nabla^h f(t\zeta)| : 0 \le t \le |x|\}.$$

But by Theorem 4.7.4(a), for fixed $\delta, 0 < \delta < \frac{1}{2}$, and $t \le |x|$,

$$\begin{aligned}|\nabla^h f(t\zeta)|^p &\le C_\delta \int_{E_\delta(t\zeta)} |\nabla^h f(y)|^p d\tau(y)\\ &\le \frac{C_\delta}{(1-t^2)^\gamma}\int_{E_\delta(t\zeta)}(1-|y|^2)^\gamma |\nabla^h f(y)|^p d\tau(y)\\ &\le \frac{C_\delta}{(1-|x|^2)^\gamma}\int_{\mathbb{B}}(1-|y|^2)^\gamma |\nabla^h f(y)|^p d\tau(y).\end{aligned}$$

From this it now follows that

$$\int_{B_{\frac{1}{2}}}(1-|x|^2)^\gamma |f(x)|^p d\tau(x) \le C\left[|f(0)|^p + \int_{\mathbb{B}}(1-|x|^2)^\gamma |\nabla^h f(x)|^p d\tau(x)\right]. \tag{10.3.3}$$

We now consider the integral over $\{\frac{1}{2} < |x| < 1\}$. Choose $\alpha > 0$ such that $\gamma - n - \alpha p > -1$. Suppose $1 < p < \infty$. Let q denote the conjugate exponent of p. Then by Hölder's inequality,

$$\begin{aligned}\left[\int_0^r \frac{|\nabla^h f(t\zeta)|}{(1-t^2)}dt\right]^p &\le \left[\int_0^r (1-t^2)^{-\alpha q - 1}dt\right]^{p/q}\int_0^r (1-t^2)^{\alpha p - 1}|\nabla^h f(t\zeta)|^p dt\\ &\le (1-r^2)^{-\alpha p}\int_0^r (1-t^2)^{\alpha p-1}|\nabla^h f(t\zeta)|^p dt.\end{aligned}$$

Hence

$$\begin{aligned}&\int_{\frac{1}{2}}^1 r^{n-1}(1-r^2)^{\gamma-n}\int_{\mathbb{S}}|f(rt) - f(0)|^p d\sigma(t)\, dr\\ &\qquad\le C\int_{\frac{1}{2}}^1 r^{n-1}(1-r^2)^{\gamma-n-\alpha p}\left[\int_0^r (1-t^2)^{\alpha p-1}M_p^p(|\nabla^h f|, t)dt\right]dr\\ &\qquad= C(I_1 + I_2),\end{aligned}$$

where $M_p^p(|\nabla^h f|, t) = \int_{\mathbb{S}} |\nabla^h f(t\zeta)|^p d\sigma(\zeta)$, and I_1 and I_2 are given by

$$\begin{aligned}I_1 &= \int_{\frac{1}{2}}^1 r^{n-1}(1-r^2)^{\gamma-n-\alpha p}\int_0^{\frac{1}{2}}(1-t^2)^{\alpha p-1}M_p^p(|\nabla^h f|, t)dt\, dr,\\ I_2 &= \int_{\frac{1}{2}}^1 r^{n-1}(1-r^2)^{\gamma-n-\alpha p}\int_{\frac{1}{2}}^r (1-t^2)^{\alpha p-1}M_p^p(|\nabla^h f|, t)dt\, dr.\end{aligned}$$

We first consider the integral I_1. As above, for all t, $0 \le t \le \frac{1}{2}$, and fixed δ, $0 < \delta < \frac{1}{2}$,

$$\begin{aligned}|\nabla^h f(t\zeta)|^p &\le \frac{C_\delta}{(1-t^2)^\gamma}\int_{E_\delta(t\zeta)}(1-|y|^2)^\gamma|\nabla^h f(y)|^p d\tau(y)\\ &\le \frac{C_\delta}{(1-t^2)^\gamma}\int_{\mathbb{B}}(1-|y|^2)^\gamma|\nabla^h f(y)|^p d\tau(y).\end{aligned}$$

Since the remaining integrals with respect to t and r are all finite,

$$I_1 \le C\int_{\mathbb{B}}(1-|y|^2)^\gamma|\nabla^h f(y)|^p d\tau(y).$$

For the integral I_2, by interchanging the order of integration, we have

$$I_2 \le \int_{\frac{1}{2}}^1 (1-t^2)^{\alpha p-1}M_p^p(|\nabla^h f|, t)\int_t^1 r^{n-1}(1-r^2)^{\gamma-n-\alpha p}dr\,dt.$$

But for $t \ge \frac{1}{2}$,

$$\int_t^1 r^{n-1}(1-r^2)^{\gamma-n-\alpha p}dr \le Ct^{n-1}(1-t^2)^{\gamma-n-\alpha p+1}.$$

Therefore,

$$\begin{aligned}I_2 &\le C\int_{\frac{1}{2}}^1 t^{n-1}(1-t^2)^{\gamma-n}\int_{\mathbb{S}}|\nabla^h f(t\zeta)|^p d\sigma(\zeta)\,dt\\ &\le C\int_{\mathbb{B}}(1-|y|^2)^\gamma|\nabla^h f(y)|^p d\tau(y).\end{aligned}$$

Thus

$$\begin{aligned}&\int_{\{\frac{1}{2}\le|y|<1\}}(1-|y|^2)^\gamma|f(y)|^p d\tau(y)\\ &\qquad\le C\left[|f(0)|^p + \int_{\mathbb{B}}(1-|y|^2)^\gamma|\nabla^h f(y)|^p d\tau(y)\right].\end{aligned}$$

Combining this with (10.3.3) proves the result for $p > 1$.

(ii) The case $0 < p \le 1$. For this part of the proof we use Theorem 8.5.1. An examination of that proof shows that the result is true whenever $|\nabla f|$ is quasi-nearly subharmonic (see also [90, Theorem 4]). Let f be $\mathcal{H}$-harmonic on $\mathbb{B}$. Then by Lemma 10.3.2 $|\nabla f_\rho|$ is quasi-nearly subharmonic[2] for all ρ, $0 < \rho < 1$. Hence as in the proof of Theorem 8.5.1,

[2] An alternate proof that $|\nabla f_\rho|$ is quasi-nearly subharmonic whenever f is $\mathcal{H}$-harmonic can be found in the two papers [63] of M. Pavlović and [17] by O. Djordjević and M. Pavlović.

$$\int_{\mathbb{S}} |f(\rho\zeta)|^p d\sigma(\zeta) \leq |f(0)|^p + C\int_{\mathbb{B}} (1-|x|)^{p-1}\rho^p|\nabla f(\rho x)|dv(x),$$

which by the change of variable $x \to \frac{x}{\rho}$

$$= |f(0)|^p + C\rho^{1-n}\int_{B_\rho} |\nabla g(x)|^p(\rho - |x|)^{p-1}dv(x).$$

Therefore, for $\gamma > n-1$,

$$\begin{aligned}
&\int_{\mathbb{B}} (1-|x|^2)^\gamma |f(x)|^p d\tau(x) \\
&\quad = n\int_0^1 \rho^{n-1}(1-\rho^2)^{\gamma-n}\int_{\mathbb{S}} |f(\rho\zeta)|^p d\sigma(\zeta)d\rho \\
&\quad \leq C_{n,\gamma}\left[|f(0)|^p + \int_0^1 (1-\rho)^{\gamma-n}\int_{B_\rho}(\rho-|x|)^{p-1}|\nabla f(x)|^p dv(x)d\rho\right],
\end{aligned}$$

which upon interchanging the order of integration

$$= C_{n,\gamma}\left[|f(0)|^p + \int_{\mathbb{B}} |\nabla f(x)|^p \int_{|x|}^1 (1-\rho)^{\gamma-n}(\rho-|x|)^{p-1}d\rho dv(x)\right].$$

But

$$\int_{|x|}^1 (1-\rho)^{\gamma-n}(\rho-|x|)^{p-1}d\rho \leq C_{n,p,\gamma}(1-|x|)^{\gamma-n+p}, \tag{10.3.4}$$

from which the result now follows. For $\gamma - n \geq 0$ the proof of inequality (10.3.4) is straightforward. When $-1 < \gamma - n < 0$, split the integral up into integrals over $[|x|, \frac{1}{2}(1+|x|)]$ and $[\frac{1}{2}(1+|x|), 1]$ respectively, and then estimate each of the integrals. □

To conclude this section we consider the analogue of Theorem 10.3.3 for eigenfunctions of Δ_h with non-zero eigenvalues. As in Definition 5.5.1, for $\lambda \in \mathbb{R}$,

$$\mathcal{H}_\lambda = \{f \in C^2(\mathbb{B}) : \Delta_h f = \lambda f\}.$$

Our first result is the following analogue of Theorem 10.3.3.

Theorem 10.3.4 *Let $f \in \mathcal{H}_\lambda, \lambda \neq 0$. Then for all p, $0 < p < \infty$ and $\gamma \in \mathbb{R}$,*

$$f \in L^p_\gamma(\tau) \iff |\nabla^h f| \in L^p_\gamma(\tau),$$

with equivalence of norms.

Proof. The proof that $f \in L^p_\gamma$ implies that $|\nabla^h f| \in L^p_\gamma$ is the same as in Theorem 10.3.3. The reverse implication follows likewise from Exercise 4.8.6. □

10.4 Integrability of Functions in $\mathcal{B}_\gamma^p$ and $\mathcal{D}_\gamma^p$

In the previous section we proved that up to constants, the spaces $\mathcal{B}_\gamma^p$ and $\mathcal{D}_\gamma^p$ are equivalent for all p, $0 < p < \infty$, whenever $\gamma > n-1$. In this section we determine the values of γ for a given p, $0 < p < \infty$, for which the spaces $\mathcal{B}_\gamma^p$ and $\mathcal{D}_\gamma^p$ are trivial. Our first theorem concerns the spaces $\mathcal{B}_\gamma^p$. However, we first prove a theorem for $\mathcal{H}$-subharmonic functions in $L_\gamma^p(\tau)$, from which our result concerning $\mathcal{B}_\gamma^p$ follows. See [85] for the analogous result for $\mathcal{M}$-subharmonic functions on the unit ball in C^m.

Theorem 10.4.1 **(a)** *Let $0 < p < \infty$. If f is a non-negative $\mathcal{H}$-subharmonic function with $f \in L_\gamma^p(\tau)$ for some $\gamma \le \min\{p(n-1),(n-1)\}$, then $f \equiv 0$.*

(b) *If $0 < p < 1$ and f is an $\mathcal{H}$-subharmonic function on $\mathbb{B}$ with $f \in L_\gamma^p(\tau)$ for some $\gamma \le \min\{p(n-1),(1-p)(n-1)\}$, then $f \equiv 0$.*

Proof. (a) Suppose $p \ge 1$ and assume $\gamma \le n-1$ and that $f \in L_\gamma^p(\tau)$ with $f \not\equiv 0$. Hence $f(a) \ne 0$ for some $a \in \mathbb{B}$. Since $f \circ \varphi_a \in L_\gamma^p(\tau)$, we can without loss of generality assume that $a = 0$. Since f^p is a non-negative $\mathcal{H}$-subharmonic function, for $0 < \rho < 1$,

$$\begin{aligned}\|f\|_{\gamma,p}^p &\ge n\int_0^\rho r^{n-1}(1-r^2)^{\gamma-n}\int_{\mathbb{S}} f(rt)^p d\sigma(t)dr\\ &\ge f(0)^p n\int_0^\rho r^{n-1}(1-r^2)^{\gamma-n}dr.\end{aligned}$$

However,

$$\lim_{\rho\to 1}\int_0^\rho r^{n-1}(1-r^2)^{\gamma-n}dr$$

is finite if and only if $\gamma - n > -1$, that is, $\gamma > n-1$. Thus $f(0) = 0$, which is a contradiction.

Suppose $0 < p < 1$. Let $f \in L_\gamma^p(\tau)$ with $f \ge 0$ and $f \not\equiv 0$. As above we can without loss of generality assume $f(0) \ne 0$. Since $f \in L_\gamma^p(\tau)$, as in (10.1.5) we have

$$f(x) \le \frac{C}{(1-|x|^2)^{\gamma/p}}\|f\|_{\gamma,p}$$

for all $x \in \mathbb{B}$. Thus for all x with $f(x) \ne 0$ we have

$$f(x)^{p-1} \ge c(1-|x|^2)^{\frac{\gamma}{p}(1-p)}$$

for a positive constant c. Therefore

$$\int_{\mathbb{S}} f(rt)^p d\sigma(t) = \int_{\mathbb{S}} f(rt)f(rt)^{p-1}d\sigma(t) \ge cf(0)(1-r^2)^{\frac{\gamma}{p}(1-p)}$$

and

$$\begin{aligned}\|f\|^p_{\gamma,p} &\geq n\int_0^\rho r^{n-1}(1-r^2)^{\gamma-n}\int_{\mathbb{S}} f(rt)^p d\sigma(t)dr\\ &\geq nf(0)\int_0^\rho r^{n-1}(1-r^2)^{\frac{\gamma}{p}-n}dr.\end{aligned}$$

However,

$$\lim_{\rho\to 1}\int_0^\rho r^{n-1}(1-r^2)^{\frac{\gamma}{p}-n}dr$$

is finite if and only if $\frac{\gamma}{p}-n>-1$, that is, $\gamma>p(n-1)$.

(b) Suppose $0<p<1$ and f is $\mathcal{H}$-subharmonic with $f\in L^p_\gamma(\tau)$ for some $\gamma\leq\min\{p(n-1),(1-p)(n-1)\}$. Let $f^+(x)=\max\{f(x),0\}$. Then f^+ is a non-negative $\mathcal{H}$-subharmonic function with $f^+\in L^p_\gamma(\tau)$ for some $\gamma\leq p(n-1)$. Thus $f^+\equiv 0$ and thus $|f|=-f$. Since $-f$ is a non-negative $\mathcal{H}$-superharmonic function on $\mathbb{B}$, by the Riesz decomposition theorem

$$|f(x)|=\int_{\mathbb{B}} G_h(x,y)d\mu(y)+\int_{\mathbb{S}} P_h(x,t)d\alpha(t),$$

where μ is a regular Borel measure on $\mathbb{B}$ with $\int_{\mathbb{B}}(1-|y|^2)^{n-1}d\mu<\infty$ and α is a non-negative finite Borel measure on $\mathbb{S}$. Since

$$P_h(x,t)\geq C(1-|x|^2)^{n-1}\quad\text{and}\quad G_h(x,y)\geq C(1-|x|^2)^{n-1}(1-|y|^2)^{n-1},$$

we have that

$$|f(x)|\geq C(1-|x|^2)^{n-1}$$

for all $x\in\mathbb{B}$. Hence, if f is not identically zero,

$$\|f\|^p_{\gamma,p}\geq C\int_0^1 r^{n-1}(1-r^2)^{\gamma-n-p(n-1)}=+\infty$$

for any γ satisfying $\gamma\leq(1-p)(n-1)$. □

As an immediate consequence of the previous theorem we have the following corollary.

Corollary 10.4.2 **(a)** *If $f\in\mathcal{B}^p_\gamma$ for some $p\geq 1$ and $\gamma\leq n-1$, then $f\equiv 0$.*
(b) *If $f\in\mathcal{B}^p_\gamma$ for some $p, 0<p<1$, and $\gamma\leq p(n-1)$, then $f\equiv 0$.*

Proof. (a) Since $|f|$ is a non-negative $\mathcal{H}$-subharmonic function, the result for $p\geq 1$ follows from part (a) of Theorem 10.4.1.

(b) Since f is $\mathcal{H}$-harmonic, both $f^+(x)=\max\{f(x),0\}$ and $f^-(x)=\max\{-f(x),0\}$ are non-negative $\mathcal{H}$-subharmonic functions on $\mathbb{B}$. Hence if

$f \in \mathcal{B}^p_\gamma$ for some $\gamma \le p(n-1)$, both f^+ and f^- are in L^p_τ with $\gamma \le p(n-1)$ and thus by (a) of Theorem 10.4.1 are identically zero. □

The proof of part (b) of Theorem 10.4.1 also proves the following.

Theorem 10.4.3 *If f is a non-negative $\mathcal{H}$-superharmonic function with $f \in L^p_\gamma$, $0 < p < 1$, $\gamma \le (1-p)(n-1)$, then $f \equiv 0$.*

For non-negative $\mathcal{H}$-harmonic functions we have the following theorem.

Theorem 10.4.4 *Let $0 < p < 1$. If h is a non-negative $\mathcal{H}$-harmonic function with $h \in \mathcal{B}^p_\gamma$ for some γ satisfying*

$$\gamma \le \max\{p(n-1), (1-p)(n-1)\},$$

then $h \equiv 0$.

Proof. If $\frac{1}{2} \le p < 1$, then $\max\{p(n-1), (1-p)(n-1)\} = p(n-1)$, and the conclusion follows from Theorem 10.4.1. Suppose $0 < p < \frac{1}{2}$, then $\max\{p(n-1), (1-p)(n-1)\} = (1-p)(n-1)$ and the conclusion follows from Theorem 10.4.3. □

Remark 10.4.5 *Theorems 10.4.1 and 10.4.3 are analogues of results proved by N. Suzuki* [92] *for Euclidean subharmonic functions on domains in $\mathbb{R}^n$. For example, Suzuki proved that if S is a non-negative (Euclidean) subharmonic function on a bounded domain D with $C^{1,1}$ boundary satisfying*

$$\int_D \frac{S^p(x)}{\delta(x)^{1+\beta(p)}} dx < \infty,$$

then $S \equiv 0$. In the above $\beta(p) = \max\{(n-1)(1-p), 0\}$ and $\delta(x)$ is the distance from x to the boundary of D. Furthermore, an example is given to show that this is the best possible.

Examples 10.4.6 In these examples we show that some of the conclusions of the previous theorems for $0 < p < 1$ are the best possible.

(i) In this example we show that for each $\gamma > (1-p)(n-1)$ there exists a non-negative $\mathcal{H}$-superharmonic function in L^p_γ. Let $f(x) = (1-|x|^2)^\alpha$. By Exercise 4.8.2 the function f is $\mathcal{H}$-superharmonic for all α, $0 < \alpha \le (n-1)$. Furthermore,

$$\int_{\mathbb{B}} (1-|x|^2)^\gamma f^p(x) d\tau(x) = n \int_0^1 r^{n-1}(1-r^2)^{\gamma-n+p\alpha} dr,$$

which is finite provided $\gamma > (n-1) - p\alpha$. Taking $\alpha = (n-1)$ gives the desired example. Thus the conclusion of Theorem 10.4.3 is the best possible.

(ii) For Theorem 10.4.3, take $f(x) = P_h(x, e_1)$. Then

$$\|f\|_{\gamma,p}^p = n\int_0^1 r^{n-1}(1-r^2)^{\gamma-n}\int_{\mathbb{S}} P_h^p(rt, e_1)d\sigma(t)dr.$$

But

$$\int_{\mathbb{S}} P_h^p(rt, e_1)d\sigma(t) = \int_{\mathbb{S}} P_h^p(re_1, t)d\sigma(t),$$

which by Corollary 5.5.8 for $\frac{1}{2} < p < 1$

$$\approx (1-r^2)^{(1-p)(n-1)}.$$

Therefore,

$$\|f\|_{\gamma,p}^p \approx \int_0^1 r^{n-1}(1-r^2)^{\gamma-n+(1-p)(n-1)}dr.$$

The above integral, however, is finite if and only if $\gamma > p(n-1)$. When $p = \frac{1}{2}$, then again by Corollary 5.5.8

$$\|f\|_{\gamma,p}^p \approx \int_0^1 r^{n-1}(1-r^2)^{\gamma-n+\frac{1}{2}(n-1)}\log\frac{1}{(1-r^2)}dr,$$

and the above integral again is finite if and only if $\gamma > \frac{1}{2}(n-1)$.

For $0 < p < \frac{1}{2}$,

$$\|f\|_{\gamma,p}^p \approx \int_0^1 r^{n-1}(1-r^2)^{\gamma-n+p(n-1)}dr,$$

and this integral is finite if and only if $\gamma > (1-p)(n-1)$. Hence for each γ not satisfying the hypothesis of Theorem 10.4.4 there exists a non-negative $\mathcal{H}$-harmonic function in $\mathcal{B}_\gamma^p$.

(iii) In this example we show that for each $\gamma > \max\{p(n-1), \frac{1}{2}(n-1)\}$ there exists a non-negative $\mathcal{H}$-subharmonic function in L_γ^p.

If $\frac{1}{2} < p < 1$, let $f(x) = P_h(x, e_1)$. Then f is a non-negative $\mathcal{H}$-subharmonic function which by the previous example is in L_γ^p whenever $\gamma > p(n-1)$.

Suppose $0 < \frac{1}{2} < p$ and $\gamma > \frac{1}{2}(n-1)$. Then we can choose $\beta > 1$ such that

$$\frac{\gamma}{p(n-1)} > \beta > \frac{1}{2p}.$$

Then $2p\beta > 1$ and $\gamma > \beta p(n-1)$. In this case take $f(x) = P_h^\beta(x, e_1)$. Then f is $\mathcal{H}$-subharmonic and

$$\int_{\mathbb{B}}(1-|x|^2)^\gamma f^p(x) = n\int_0^1 r^{n-1}(1-r^2)^{\gamma-n}\int_{\mathbb{S}} P_h^{\beta p}(re_1, t)d\sigma(t),$$

which by Corollary 5.5.8, since $\beta p > \frac{1}{2}$,

$$\approx \int_0^1 r^{n-1}(1-r^2)^{\gamma-n+(1-\beta p)(n-1)}dr.$$

The above integral is finite provided $\gamma > \beta p(n-1)$.

Remark 10.4.7 *As mentioned in Remark 10.4.5, N. Suzuki proved that for* $0 < p < 1$ *and* $\gamma > p(n-1)$*, there exists a non-negative Euclidean subharmonic function* f *on* $\mathbb{B}$ *such that* $f \in L^p_\gamma(\tau)$*. Since the two concepts coincide when* $n = 2$*, we provide a short proof of the example given by the author in* [85] *in the following theorem. This still leaves the question of the existence of a non-negative* $\mathcal{H}$*-subharmonic function* f *on* $\mathbb{B}$ *with* $f \in L^p_\gamma(\tau)$*,* $\gamma > p(n-1)$*, when* $n \geq 3$.

Theorem 10.4.8 [85] *If* $n = 2$ *and* $0 < p < 1$*, then for each* $\gamma > p$ *there exists a non-negative subharmonic function* $f_\gamma \in L^p_\gamma(\tau)$.

Proof. For $0 < \beta < \pi/2$, let S_β be the angular region with vertex at 1 defined by

$$S_\beta = \{z \in \mathbb{D} : |\arg(1-z)| < \beta,\ |1-z| < \cos\beta\}.$$

Let φ_β be a conformal mapping of S_β onto $\mathbb{D}$ mapping the boundary of S_β onto the boundary of $\mathbb{D}$ with $\varphi_\beta(1) = 1$. Let g_β be the function defined on $\mathbb{D}$ by

$$g_\beta(z) = \begin{cases} P_h(\varphi_\beta(z), 1), & z \in S_\beta \\ 0, & z \in \mathbb{D} \setminus S_\beta. \end{cases}$$

Then the function g_β is subharmonic on $\mathbb{D}$. By [55, Lemma 2.2] there exists a non-zero analytic function h defined on a neighborhood N of 1 such that

$$1 - \varphi_\beta(z) = (1-z)^b h(z)$$

for all $z \in N \cap S_\beta$, where $b = \pi/(2\beta)$. Set $b = 1 + \epsilon(\beta)$ where $\epsilon(\beta) \to 0$ as $\beta \to \pi/2$. Then, for all $z \in S_\beta$,

$$g_\beta(z) \leq P_h(\varphi_\beta(z), 1) \leq \frac{C}{|1-z|^{1+\epsilon(\beta)}}.$$

Hence,

$$\begin{aligned}\int_{\mathbb{D}} (1-|z|^2)^\gamma g_\beta^p(z)d\tau(z) &\leq C\int_0^1 (1-r^2)^{\gamma-2}\int_0^{2\pi} \frac{\chi_{S_\beta}(re^{i\theta})}{|1-re^{i\theta}|^{p+p\epsilon(\beta)}}d\theta\, r\,dr \\ &\leq C\int_0^1 (1-r^2)^{\gamma-p-p\epsilon(\beta)-1} r\,dr.\end{aligned}$$

If $\gamma > p$ we can choose β sufficiently close to $\pi/2$ such that $\gamma - p - p\epsilon(\beta) > 0$, in which case the above integral is finite. For such a β, define $f_\gamma = g_\beta$. □

We now consider the spaces $\mathcal{D}_\gamma^p$, $0 < p < \infty$, $\gamma \le n-1$. By Exercise 4.8.1,

$$|x|^2|\nabla^h f(x)|^2 = (1-|x|^2)^2|\langle x, \nabla f(x)\rangle|^2 + (1-|x|^2)^2 \sum_{i<j} |T_{i,j}f(x)|^2, \quad (10.4.1)$$

where $T_{i,j}f(x) = x_i(\partial f/\partial x_j) - x_j(\partial f/\partial x_i)$. Thus $f \in \mathcal{D}_\gamma^p$ if and only if $|\langle x, \nabla f(x)\rangle|$ and $|T_{i,j}f|$ are in $L_{\gamma+p}^p$ for all i,j. This leads to the following theorem.

Theorem 10.4.9 **(a)** *If $f \in \mathcal{D}_\gamma^p$ for some $p \ge 1$ and $\gamma \le (n-1) - p$, then f is constant.*

(b) *If $f \in \mathcal{D}_\gamma^p$ for some p, $0 < p < 1$, and $\gamma \le p(n-1) - p$, then f is constant.*

Proof. Since $T_{i,j}f$ is $\mathcal{H}$-harmonic whenever f is $\mathcal{H}$-harmonic, if $f \in \mathcal{D}_\gamma^p$, $0 < p < \infty$, then $T_{i,j}f \in \mathcal{B}_{\gamma+p}^p$ for all i,j. Hence if f satisfies either (a) or (b), then by Corollary 10.4.2 $T_{i,j}f = 0$ for all i,j.

To conclude the proof we show that this implies that f is constant. If $T_{i,j}f(x) = 0$ for all i,j and all $x \in \mathbb{B}$, then by (10.4.1),

$$|\langle x, \nabla f(x)\rangle| = |x||\nabla f(x)|,$$

and as a consequence $\nabla f(x) = \lambda(x)\,x$ for some scalar function λ. We now show that this implies that f is a radial function.

Consider the sphere S_r of radius r and let p_1 and p_2 be any two points on S_r. Also let $\phi(t)$, $0 \le t \le 1$, be any C^1 curve in S_r from p_1 to p_2. Since $|\phi(t)| = r$, by differentiation it follows that $\langle \phi'(t), \phi(t)\rangle = 0$. Thus

$$\begin{aligned} f(p_2) - f(p_1) &= \int_0^1 \langle \nabla f(\phi(t)), \phi'(t)\rangle dt \\ &= \int_0^1 \lambda(\phi(t))\langle \phi(t), \phi'(t)\rangle dt = 0. \end{aligned}$$

Thus $f(p_2) = f(p_1)$; that is, f is a radial function. However, any radial $\mathcal{H}$-harmonic function is a constant. To see this, if f is radial and satisfies $\Delta_h f = 0$, then either f is constant or f is a constant multiple of the Green's function g_h, which is not $\mathcal{H}$-harmonic on $\mathbb{B}$. □

Example 10.4.10 **(i)** Suppose f is $\mathcal{H}$-harmonic on $\mathbb{B}$ such that $|\nabla f(x)| \le C$. Then $|\nabla^h f(x)| \le C(1-|x|^2)$ and

$$\int_{\mathbb{B}} (1-|x|^2)^\gamma |\nabla^h f(x)|^p d\tau(x) \le C \int_0^1 r^{n-1}(1-r^2)^{\gamma-n-p} dr,$$

which is finite if and only if $\gamma > (n-1) - p$.

If the dimension of $\mathbb{R}^n$ is even, then as in Example 6.1.3, $P_h[q](x)$ is a polynomial whenever q is a polynomial on $\mathbb{S}$. Thus $|\nabla P_h[q](x)| \le C$ for all $x \in \mathbb{B}$.

If the dimension of $\mathbb{R}^n$ is odd, it is still possible to find a function f which is $\mathcal{H}$-harmonic on $\mathbb{B}$ for which $|\nabla f| \le C$ on $\mathbb{B}$. We illustrate this with $n = 3$. If as above we let $q(t) = t_1^2$, then by Example 6.1.3

$$P_h[q](x) = \tfrac{1}{3} + (x_1^2 - \tfrac{1}{3}|x|^2) S_{3,2}(|x|),$$

where

$$S_{3,2}(r) = c_{3,2} F(2, -\tfrac{1}{2}; \tfrac{7}{2}; r^2).$$

Using the fact that ([1])

$$\frac{d}{dz} F(a,b;c;z) = \frac{ab}{c} F(a+1, b+1; c+1; z),$$

and that $F(a,b;c;z)$ converges absolutely on $|z| = 1$ whenever $c - a - b > 0$, we have that

$$S'_{3,2}(r) = c\, r F(3, \tfrac{1}{2}; \tfrac{9}{2}; r^2),$$

and the series converges absolutely for all r, $0 \le r \le 1$. Thus $|\nabla P_h[q](x)|$ is bounded on $\mathbb{B}$. Note, even though $P_h[q](x)$ is C^1 on $\overline{\mathbb{B}}$, it is not C^2. The function $S''_{3,2}$ does not converge absolutely on $|z| = 1$. For $n > 3$, the result follows by Exercise 6.4.2.

(ii) Suppose $n = 2$ and u is harmonic on $\mathbb{D}$. If we let $u = \operatorname{Re} f$ where f is analytic on $\mathbb{D}$, then $|\nabla u| = \sqrt{2}|f'(z)|$. Thus $u \in \mathcal{D}^p_\gamma$ if and only if

$$\int_{\mathbb{D}} (1-|z|^2)^{\gamma-2+p} |f'(z)|^p dA(z) < \infty. \tag{10.4.2}$$

However, since $|f'(z)|^p$ is subharmonic for all $p > 0$,

$$\begin{aligned}\int_{R<|z|<1} (1-|z|^2)^{\gamma-2+p} |f'(z)|^p dA(z) &\ge M_p^p(f', R) \int_R^1 r(1-r^2)^{\gamma-2+p} dr \\ &\ge C_{\gamma,p}(1-R^2)^{\gamma+p-1}.\end{aligned}$$

Therefore,

$$M_p^p(f', R) \le C(1-R^2)^{-\gamma-p+1} \int_{R<|z|<1} (1-|z|^2)^{\gamma-2+p} |f'(z)|^p d\tau(z).$$

Hence if $f \in \mathcal{D}_\gamma^p$ for $\gamma \le 1-p$, $\lim_{R\to 1} M_p^p(f', R) = 0$. Thus f is constant on $\mathbb{B}$. Therefore when $n = 2$, if $u \in \mathcal{D}_\gamma^p$ for $\gamma \le (1-p)$, then u is constant. Thus when $n = 2$ the result of Theorem 11.8.2 is not best possible. Also, for $n = 2$, $f(z) = z$ is in $\mathcal{D}_\gamma^p$ for all $\gamma > (1-p)$.

(iii) For $p > 0$ and $\gamma \in \mathbb{R}$, set

$$D_\gamma^p = \{f : f \text{ is Euclidean harmonic with } |\nabla^h f| \in L_\gamma^p(\tau)\}.$$

Then $f \in D_\gamma^p$ if and only if

$$\int_{\mathbb{B}} (1-|x|^2)^{\gamma-n+p} |\nabla f(x)|^p dv(x) < \infty.$$

By [76, Theorem 4.1.4], when $n \ge 3$, $|\nabla f|^p$ is subharmonic when $p \ge (n-2)/(n-1)$. Hence as in Example (ii), for $p \ge (n-2)/(n-1)$,

$$M_p^p(|\nabla f|, R) \le C_{\gamma,p}(1-R^2)^{n-\gamma-p-1} \int_{R<|x|<1} (1-|x|^2)^{\gamma-n+p} |\nabla f(x)|^p dv(x).$$

Thus if $\gamma \le (n-1) - p$ we have

$$\lim_{R\to 1} M_p^p(|\nabla f|, R) = 0$$

from which it follows that f is constant on $\mathbb{B}$. The function $f(x) = x_1^2 - x_2^2$ is harmonic on $\mathbb{B}$ with $|\nabla f(x)| \le 2$. Thus $f \in D_\gamma^p$ for any $\gamma > (n-1) - p$.

Remark 10.4.11 *In view of the above examples it is conjectured that if $f \in \mathcal{D}_\gamma^p$, $0 < p < \infty$, for any $\gamma \le (n-1) - p$, then f is constant.*

10.5 Integrability of Eigenfunctions of Δ_h

In this section we investigate the integrability of eigenfunctions of Δ_h with real non-zero eigenvalues λ. If λ and α are related by

$$\lambda = 4(n-1)^2\alpha(\alpha-1), \qquad \alpha \ne 0,\ 1, \tag{10.5.1}$$

then each $P_h^\alpha(x,t)$, $t \in \mathbb{S}$, is an eigenfunction of Δ_h with eigenvalue λ. Solving (10.5.1) for α gives

$$\alpha = \frac{1}{2} \pm \frac{1}{2(n-1)}\sqrt{\lambda + (n-1)^2}. \tag{10.5.2}$$

Thus λ and α are real if and only if $\lambda \geq -(n-1)^2$. Our first theorem is the following.

Theorem 10.5.1 *Let $0 < p < \infty$ and suppose λ is real with $\lambda \geq -(n-1)^2$ and $\lambda \neq 0$. If f is a radial function on $\mathbb{B}$ with $f(0) \neq 0$, then*

$$f \in L^p_\gamma(\tau) \cap \mathcal{H}_\lambda$$

if and only if $\gamma > (n-1) + \frac{p}{2}\left(\sqrt{\lambda + (n-1)^2} - (n-1)\right)$.

Proof. If $f \in \mathcal{H}_\lambda$ is radial, then by Theorem 5.5.3,

$$f(x) = f(0)g_\alpha(x),$$

where λ and α are related by (10.5.1) and g_α is given by (5.5.4). Hence with $\alpha = \frac{1}{2} + \frac{1}{2(n-1)}\sqrt{\lambda + (n-1)^2}$,

$$\int_{\mathbb{B}} (1-|x|^2)^\gamma f^p(x)d\tau(x) = f^p(0)\int_{\mathbb{B}} (1-|x|^2)^{\gamma-n} g^p_\alpha(x)d\nu(x). \quad (10.5.3)$$

Since $\alpha > \frac{1}{2}$, by Corollary 5.5.8 $g_\alpha(x)^p \approx (1-|x|^2)^{p(1-\alpha)(n-1)}$. Thus the integral in (10.5.3) is finite if and only if $\gamma > (n-1) - p(1-\alpha)(n-1)$ or

$$\gamma > (n-1) + \frac{p}{2}\left(\sqrt{(n-1)^2 + \lambda} - (n-1)\right).$$

□

Theorem 10.5.2 *Let $0 < p < \infty$ and suppose λ is real with $\lambda \geq -(n-1)^2$. Then*

$$L^p_\gamma(\tau) \cap \mathcal{H}_\lambda = \{0\}$$

(a) *for all $\gamma \leq (n-1) + \frac{p}{2}\left(\sqrt{(n-1)^2 + \lambda} - (n-1)\right)$ when $p \geq 1$, and*

(b) *for all $\gamma \leq \frac{p}{2}\left((n-1) + \sqrt{(n-1)^2 + \lambda}\right)$ when $0 < p < 1$.*

Proof. (a) Suppose $p \geq 1$ and $f \in L^p_\gamma \cap \mathcal{H}_\lambda$. Then $f \circ \varphi_a \in L^p_\gamma$ for all $a \in \mathbb{B}$. Hence by integration in polar coordinates

$$\|f \circ \varphi_a\|^p_{\gamma,p} = n\int_0^1 r^{n-1}(1-r^2)^{\gamma-n}\int_{\mathbb{S}} |f(\varphi_a(r\zeta))|^p d\sigma(\zeta)dr,$$

which by Hölder's inequality

$$\geq n\int_0^1 r^{n-1}(1-r^2)^{\gamma-n}\left|\int_{\mathbb{S}} f(\varphi_a(rt))d\sigma(t)\right|^p dr.$$

But by Theorem 5.5.5, since $f \in \mathcal{H}_\lambda$,

$$\int_{\mathbb{S}} f(\varphi_a(rt))d\sigma(t) = g_\alpha(r)f(a).$$

Therefore with $\alpha = \frac{1}{2} + \frac{1}{2(n-1)}\sqrt{(n-1)^2 + \lambda}$,

$$\begin{aligned} \|f \circ \varphi_a\|_{\gamma,p}^p &\geq n|f(a)|^p \int_0^1 r^{n-1}(1-r^2)^{\gamma-n} g_\alpha^p(r)dr \\ &\geq C|f(a)|^p \int_0^1 r^{n-1}(1-r^2)^{\gamma-n+p(1-\alpha)(n-1)}dr. \end{aligned}$$

The above integral is finite, however, if and only if

$$\gamma > (n-1) - p(1-\alpha)(n-1) = (n-1) + \frac{p}{2}\left(\sqrt{(n-1)^2+\lambda} - (n-1)\right).$$

Thus if $\|f\circ\varphi_a\|_{\gamma,p} < \infty$ for some $\gamma \leq (n-1)+\frac{p}{2}\left(\sqrt{(n-1)^2+\lambda} - (n-1)\right)$, we must have $f(a) = 0$, hence the result. Taking $\alpha' = \frac{1}{2} - \frac{1}{2(n-1)}\sqrt{(n-1)^2+\lambda}$ gives the same result since $(1-\alpha') = \alpha$. If $\alpha = \frac{1}{2}$ $(\lambda = -(n-1)^2)$, then

$$\|f \circ \varphi_a\|_{\gamma,p}^p \geq C|f(a)|^p \int_0^1 r^{n-1}(1-r^2)^{\gamma-n-\frac{p}{2}(n-1)}\left(\log\frac{1}{(1-r^2)}\right)^p dr,$$

and the above integral again is finite if and only if $\gamma > (n-1) - \frac{p}{2}(n-1)$.

(b) Suppose now that $0 < p < 1$ and $f \in L^p_\gamma \cap \mathcal{H}_\lambda$. Let $y \in \mathbb{B}$. Then since $|f|$ is quasi-nearly $\mathcal{H}$-subharmonic, for $0 < \delta < \frac{1}{2}$,

$$|f(y)|^p \leq C_\delta \int_{E_\delta(y)} |f(x)|^p d\tau(x),$$

which since $(1-|x|^2) \approx (1-|y|^2)$ for all $x \in E_\delta(y)$,

$$\begin{aligned} &\leq \frac{C_\delta}{(1-|y|^2)^\gamma}\int_{E_\delta(y)} (1-|x|^2)^\gamma |f(x)|^p d\tau(x) \\ &\leq \frac{C_\delta}{(1-|y|^2)^\gamma}\|f\|_{\gamma,p}^p. \end{aligned}$$

Thus for all y with $f(y) \neq 0$,

$$|f(y)|^{p-1} \geq C(1-|y|^2)^{\frac{\gamma}{p}(1-p)}.$$

Therefore,

$$\begin{aligned}
\int_{\mathbb{S}} |f(rt)|^p d\sigma(t) &= \int_{\mathbb{S}} |f(rt)||f(rt)|^{p-1} d\sigma(t) \\
&\geq C(1-r^2)^{\frac{\gamma}{p}(1-p)} \int_{\mathbb{S}} |f(rt)| d\sigma(t) \\
&\geq C(1-r^2)^{\frac{\gamma}{p}-\gamma} |f(0)| g_\alpha(r).
\end{aligned}$$

Therefore, if $\lambda > -(n-1)^2$ and $\alpha = \frac{1}{2} + \frac{1}{2(n-1)}\sqrt{(n-1)^2+\lambda}$, then

$$\|f\|_{\gamma,p}^p \geq C|f(0)| \int_0^1 r^{n-1}(1-r^2)^{\frac{\gamma}{p}-n+(1-\alpha)(n-1)} dr.$$

The above integral is finite if and only if

$$\frac{\gamma}{p} > (n-1) - (1-\alpha)(n-1) = \frac{1}{2}\left((n-1) + \sqrt{(n-10^2+\lambda}\right),$$

from which the result follows. If $\lambda = -(n-1)^2$, that is, $\alpha = \frac{1}{2}$, then the result follows as in (a). □

Example 10.5.3 In this example we show that the result of Theorem 10.5.2 is sharp for all

$$p \geq \frac{(n-1)}{\left((n-1) + \sqrt{(n-1)^2+\lambda}\right)}, \qquad \lambda \geq -(n-1)^2.$$

(i) Suppose first that $p \geq 1$. Then with $\alpha = \frac{1}{2} + \frac{1}{2(n-1)}\sqrt{(n-1)^2+\lambda}$, by Theorem 10.5.1, the function $g_\alpha \in L_\gamma^p$ if and only if

$$\gamma > (n-1) + \frac{p}{2}\left(\sqrt{(n-1)^2+\lambda} - (n-1)\right).$$

(ii) Suppose $0 < p < 1$. Consider the function $P_h^\alpha(x, e_1)$ with $\alpha = \frac{1}{2} + \frac{1}{2(n-1)}\sqrt{(n-1)^2+\lambda}$. Then,

$$\begin{aligned}
\|P_h^\alpha\|_{\gamma,p}^p &= n\int_0^1 r^{n-1}(1-r^2)^{\gamma-n} \int_{\mathbb{S}} P_h^{p\alpha}(rt, e_1) d\sigma(t)\, dr \\
&= n\int_0^1 r^{n-1}(1-r^2)^{\gamma-n} g_{p\alpha}(r) dr.
\end{aligned}$$

If $p\alpha > \frac{1}{2}$, then $g_{p\alpha}(r) \approx (1-r^2)^{(1-p\alpha)(n-1)}$, and the above integral is finite if and only if

$$\gamma > p\alpha(n-1) = \frac{p}{2}\left((n-1) + \sqrt{(n-1)^2+\lambda}\right).$$

Likewise, if $p\alpha = \frac{1}{2}$, $g_{p\alpha}(r) \approx (1-r^2)^{\frac{1}{2}(n-1)} \log \frac{1}{(1-r^2)}$, and the integral again is finite if and only if

$$\gamma > \frac{1}{2}(n-1) = \frac{p}{2}\left((n-1) + \sqrt{(n-1)^2 + \lambda}\right).$$

But $p\alpha \geq \frac{1}{2}$ if and only if

$$p \geq \frac{(n-1)}{(n-1) + \sqrt{(n-1)^2 + \lambda}}.$$

Whether the conclusion of Theorem 10.5.2 is the best possible for the values of p satisfying $0 < p < (n-1)/((n-1) + \sqrt{(n-1)^2 + \lambda})$ is not known to the author. As the following theorem proves, for non-negative functions in $\mathcal{H}_\lambda$ it is not. For $\lambda = 0$ this is Theorem 10.4.3.

Theorem 10.5.4 *Let $0 < p < 1$ and let λ be real with $\lambda > -(n-1)^2$. If f is a non-negative function in $\mathcal{H}_\lambda \cap L^p_\gamma(\tau)$ for some γ satisfying*

$$\gamma \leq \max\left\{\tfrac{1}{2}p\left((n-1) + \sqrt{(n-1)^2+\lambda}\right),\right.$$
$$\left.(n-1) - \tfrac{1}{2}p\left((n-1) + \sqrt{(n-1)^2+\lambda}\right)\right\},$$

then $f \equiv 0$ on $\mathbb{B}$. The result is the best possible for all p, $0 < p < 1$.

Proof. If $f \in L^p_\gamma \cap \mathcal{H}_\lambda$ for some $\lambda \leq \frac{1}{2}p((n-1) + \sqrt{(n-1)^2+\lambda})$, then by Theorem 10.5.2 $f \equiv 0$ on $\mathbb{B}$. On the other hand, if $f \in \mathcal{H}_\lambda$ is non-negative, then by Theorem 5.5.2,

$$f(x) = \int_{\mathbb{S}} P^\alpha_h(x,t)\,d\mu(t)$$

for some non-negative measure μ on $\mathbb{S}$ and $\alpha = \frac{1}{2} + \frac{1}{2(n-1)}\sqrt{(n-1)^2+\lambda}$. Since $\alpha > 0$ we have $P^\alpha_h(x,t) \geq C(1-|x|^2)^{\alpha(n-1)}$, and thus

$$\int_{\mathbb{B}} (1-|x|^2)^\gamma |f(x)|^p d\tau(x) \geq C[\mu(\mathbb{S})]^p \int_0^1 r^{n-1}(1-r^2)^{\gamma-n+\alpha p(n-1)}dr.$$

But the integral on the right is finite if and only if

$$\gamma > (n-1) - \alpha p(n-1) = (n-1) - \tfrac{1}{2}p\left((n-1) + \sqrt{(n-1)^2+\lambda}\right).$$

Hence if $f \in L^p_\gamma$ for some $\gamma \leq (n-1) - \frac{1}{2}p((n-1) + \sqrt{(n-1)^2+\lambda})$, we have $f \equiv 0$. The function $P^\alpha_h(x, e_1)$ shows that the result is the best possible. □

Our final result of this section concerns the space $\mathcal{H}^p_\lambda$. Recall that for $0 < p < \infty$,

$$\mathcal{H}^p_\lambda = \left\{ f \in \mathcal{H}_\lambda : \sup_{0<r<1} \int_{\mathbb{S}} |f(rt)|^p d\sigma(t) < \infty \right\}.$$

If $f \in \mathcal{H}_\lambda$, $\lambda \geq 0$, then we have that $|f|$ is $\mathcal{H}$-subharmonic on $\mathbb{B}$.

Theorem 10.5.5 [89] *Let* $1 \leq p < \infty$. *If* $f \in \mathcal{H}^p_\lambda$ *for some* $\lambda > 0$, *then* $f \equiv 0$.

Proof. Since $\mathcal{H}^p_\lambda \subset \mathcal{H}^1_\lambda$ for all $p \geq 1$, it suffices to prove the result for $p = 1$. Without loss of generality we assume f is real valued, and for $\epsilon > 0$, set $f_\epsilon = f + i\epsilon$. Then $|f_\epsilon|$ is C^2 on $\mathbb{B}$. Although $f_\epsilon \notin \mathcal{H}^1_\lambda$, the function $|f_\epsilon|$ is $\mathcal{H}$-subharmonic with least $\mathcal{H}$-harmonic majorant $F_{|f_\epsilon|}$ satisfying $F_{|f_\epsilon|} \leq F_{|f|} + \epsilon$ where $F_{|f|}$ is the least $\mathcal{H}$-harmonic majorant of $|f|$. Since $\Delta_h f = \lambda f$,

$$\Delta_h |f_\epsilon| = \epsilon^2 |f_\epsilon|^{-3} |\nabla^h f|^2 + \lambda |f_\epsilon|^{-1} |f|^2 \geq \lambda |f_\epsilon|^{-1} |f|^2.$$

Since $|f_\epsilon|$ is C^2 and has an $\mathcal{H}$-harmonic majorant, we have by Theorem 9.2.4 and the above that

$$\begin{aligned}\int_{\mathbb{B}} &(1 - |y|^2)^{n-1} |f_\epsilon(y)|^{-1} |f(y)|^2 d\tau(y) \\ &\leq \lambda^{-1} \int_{\mathbb{B}} (1 - |y|^2)^{n-1} \Delta_h |f_\epsilon(y)| d\tau(y) \leq C F_{|f_\epsilon|}(0) \leq C,\end{aligned}$$

where C is a constant independent of ϵ. Hence by Fatou's lemma,

$$\int_{\mathbb{B}} (1 - |y|^2)^{n-1} |f(y)| d\tau(y) < \infty.$$

Thus $f \in L^1_\gamma(\tau) \cap H_\lambda$, $\lambda > 0$ with $\gamma = (n-1)$, which is strictly less than $\frac{1}{2}[(n-1) + \sqrt{(n-1)^2 + \lambda}]$. Thus by Theorem 10.5.2, $f \equiv 0$. □

Example 10.5.6 The conclusion of the previous theorem is false when $0 < p < 1$. The function $P^\alpha_h(x, e_1)$ is an eigenfunction of Δ_h with eigenvalue $\lambda_\alpha = 4(n-1)^2 \alpha(\alpha - 1)$. It is well known that $P_h \in \mathcal{H}^p$ for all p, $0 < p < 1$. Hence if $\alpha > 1$, $P_h \in \mathcal{H}^p_{\lambda_\alpha}$ for all p, $0 < p \leq \frac{1}{\alpha}$.

10.6 Three Theorems of Hardy and Littlewood

In this section we prove three analogues of well-known results of Hardy and Littlewood concerning the means $M_p(f, r)$ of analytic functions in the unit disc. Our first result concerns the comparative rate of growth of $M_p(f,r)$ and $M_p(f',r)$, where for $0 < p < \infty$ and $0 < r < 1$,

$$M_p(f,r) = \left[\int_{\mathbb{S}} |f(rt)|^p d\sigma(t)\right]^{1/p}.$$

As usual $M_\infty(f,r) = \sup\{|f(rt)| : t \in \mathbb{S}\}$. When $n = 2$, proofs of the results for analytic functions in the unit disc may be found in [19]. The classical result of Hardy and Littlewood [34] on the unit disc $\mathbb{D}$ is as follows. If f is analytic in $\mathbb{D}$, $0 < p < \infty$, and $\alpha > 0$, then

$$M_p(f,r) = O\left((1-r)^{-\alpha}\right) \iff M_p(f',r) = O\left((1-r)^{-(\alpha+1)}\right).$$

The proofs of the extension to $\mathcal{H}$-harmonic functions on $\mathbb{B}$ and the invariant gradient ∇^h will rely heavily on the fact that $|f|^p$ and $|\nabla^h f|^p$ are quasi-nearly $\mathcal{H}$-subharmonic for all p, $0 < p < \infty$.

Theorem 10.6.1 *If f is $\mathcal{H}$-harmonic on $\mathbb{B}$, $0 < p \le \infty$, and $\alpha > 0$, then*

$$M_p(f,r) = O\left((1-r^2)^{-\alpha}\right) \iff M_p(|\nabla^h f|,r) = O\left((1-r^2)^{-\alpha}\right)$$

as $r \to 1$.

This result was originally proved by M. Pavlović [62] for $\mathcal{M}$-harmonic functions on the unit ball in $\mathbb{C}^m$ (see also [84, Theorem 10.7]).

Proof. For the proof we will assume that $0 < p < \infty$. The case $p = \infty$ follows similarly. Suppose $M_p(f,r) = O((1-r^2)^{-\alpha})$. Then by Theorem 4.7.4(b), for $A \in O(n)$,

$$|\nabla^h f(rAe_1)|^p \le C_{\delta,p} \int_{E_\delta(rAe_1)} |f(y)|^p d\tau(y),$$

which by the invariance of τ

$$\le C_{\delta,p} \int_{E_\delta(re_1)} |f(Ay)|^p d\tau(y).$$

Thus

$$\begin{aligned} M_p^p(|\nabla^h f|,r) &= \int_{O(n)} |\nabla^h f(rAe_1)|^p dA \\ &\le C_{\delta,p} \int_{E_\delta(re_1)} \int_{O(n)} |f(Ay)|^p dA\, d\tau(y). \end{aligned}$$

But

$$\int_{O(n)} |f(Ay)|^p dA = O\left((1-|y|^2)^{-\alpha}\right).$$

But for $y \in E(re_1, \delta)$, $(1 - |y|^2) \approx (1 - r^2)$. Therefore

$$M_p^p(|\nabla^h f|, r) = O\left((1 - r^2)^{-\alpha}\right).$$

Conversely, suppose $M_p^p(|\nabla^h f|, r) = O\left((1 - r^2)^{-\alpha}\right)$. Without loss of generality we assume $f(0) = 0$. By the fundamental theorem of calculus,

$$|f(r\zeta)| \le \int_0^r \frac{|\nabla^h f(s\zeta)|}{(1 - s^2)} ds.$$

For $j = 0, 1, 2, \dots$, set $r_j = 1 - 2^{-j}$. Also, for $0 < r < 1$, let $m \in \mathbb{N}$ be such that $r_{m-1} < r \le r_m$. Then

$$\begin{aligned} |f(r\zeta)| &\le \sum_{j=0}^{m-1} \int_{r_j}^{r_{j+1}} \frac{|\nabla^h f(s\zeta)|}{(1 - s^2)} ds \\ &\le C \sum_{j=0}^{m-1} \sup\{|\nabla^h f(s\zeta)| : s \in (r_j, r_{j+1})\}. \end{aligned}$$

If $r_j < s < r_{j+1}$, then by (2.1.7)

$$\begin{aligned} |\varphi_{r_j\zeta}(s\zeta)|^2 &= \frac{(1 - r_j s)^2 - (1 - r_j^2)(1 - s^2)}{(1 - r_j s)^2} \\ &\le \frac{(s - r_j)^2}{(1 - r_j)^2} \le \frac{1}{4}. \end{aligned}$$

Thus $s\zeta \in E(r_j\zeta, \frac{1}{2})$ for all s, $r_j < s < r_{j+1}$. As a consequence, $E(s\zeta, \frac{1}{4}) \subset E(r_j\zeta, \frac{3}{4})$ for all such ζ. Since $|\nabla^h f|$ is quasi-nearly $\mathcal{H}$-subharmonic,

$$\sup\{|\nabla^h f(s\zeta)|^p : r_j < s < r_{j+1}\} \le C_p \int_{E(r_j\zeta, \frac{3}{4})} |\nabla^h f(y)|^p d\tau(y).$$

Therefore

$$\begin{aligned} &\int_{O(n)} \sup\{|\nabla^h f(sAe_1)|^p : r_j < s < r_{j+1}\} dA \\ &\le C_p \int_{E_{\frac{3}{4}}(r_j e_1)} \int_{O(n)} |\nabla^h f(Ay)|^p dA d\tau(y) \\ &= C_p \int_{E_{\frac{3}{4}}(r_j e_1)} M_p^p(|\nabla^h f|, |y|) d\tau(y). \end{aligned}$$

But for $y \in E(r_j e_1, \frac{3}{4})$,

$$M_p^p(|\nabla^h f|, |y|) \le C \frac{1}{(1 - r_j^2)^{p\alpha}} \le C\left(2^j\right)^{p\alpha}.$$

Therefore,

$$\int_{\mathbb{S}} \sup\{|\nabla^h f(s\zeta)|^p : s \in (r_j, r_{j+1})\} d\sigma(\zeta) \le C(2^{p\alpha})^j.$$

Thus if $0 < p < 1$,

$$\begin{aligned}\int_{\mathbb{S}} |f(r\zeta)|^p d\sigma(\zeta) &\le \sum_{j=0}^{m-1} \int_{\mathbb{S}} \sup\{|\nabla^h f(s\zeta)|^p : s \in (r_j, r_{j+1})\} d\sigma(\zeta)\\ &\le C \sum_{j=0}^{m-1} (2^{p\alpha})^j \le C_{p,\alpha}(2^m)^{p\alpha}\\ &\le C_{p,\alpha}(1-r^2)^{-p\alpha}.\end{aligned}$$

On the other hand, for $1 \le p < \infty$, by Minkowski's inequality,

$$\begin{aligned}\left[\int_{\mathbb{S}} |f(r\zeta)|^p d\sigma(\zeta)\right]^{1/p} &\le \sum_{j=0}^{m-1} \left[\int_{\mathbb{S}} \sup\{|\nabla^h f(s\zeta)|^p : s \in (r_j, r_{j+1})\} d\sigma(\zeta)\right]^{1/p}\\ &\le C \sum_{j=0}^{m-1} (2^{\alpha})^j \le C_\alpha (2^m)^\alpha\\ &\le C_\alpha (1-r^2)^{-\alpha}.\end{aligned}$$

Hence, in either case, $M_p(f, r) \le C(1-r^2)^{-\alpha}$. □

Our second theorem is an application of the previous one.

Theorem 10.6.2 [34] *Let f be $\mathcal{H}$-harmonic on $\mathbb{B}$ and suppose*

$$M_p(f, r) \le \frac{C}{(1-r)^\beta}, \quad 0 < p < \infty,\ \beta \ge 0.$$

Then there exists a constant K, depending only on p, β, and n, such that

$$M_q(f, r) \le \frac{KC}{(1-r)^{\beta + (n-1)(\frac{1}{p} - \frac{1}{q})}}, \quad 0 < p < q \le \infty.$$

Proof. We first prove the result for $q = \infty$. Our proof will use the notation of Theorem 8.5.1. Let $0 < r < 1$. Set $r_j = 1 - \frac{1}{2^j}$, $j = 0, 1, 2, \ldots, N$, where N is the first integer such that $r_N > r$. Then as in Theorem 8.5.1,

$$|f(r\zeta)| \le |f(0)| + \log 2 \sum_{k=1}^{N} \sup_{t \in [r_{k-1}, r_k)} |\nabla^h f(t\zeta)|.$$

For $x \in \mathbb{B}$ let $B(x) = B(x, \frac{1}{4}(1-|x|^2))$, and for $j = 2, 3, 4, \ldots, N$, set

$$A_j = \{x \in \mathbb{B} : r_{j-2} < |x| < r_{j+2}\}.$$

Then as in the proof of Theorem 8.5.1

$$\sup_{t\in[r_{k-1},r_k)} |\nabla^h f(t\zeta)| \le C_n \left(\int_{A_k} |\nabla^h f(x)|^p d\tau(x)\right)^{1/p}.$$

Suppose $M_p(f,r) \le C(1-r)^{-\beta}$. Then by Theorem 10.3.4, $M_p(|\nabla^h f|, r) \le C(1-r)^{-\beta}$. Note, an examination of the proof of Theorem 10.3.4 shows that this part of the result is still valid when $\beta = 0$. Therefore,

$$\begin{aligned}\int_{A_k} |\nabla^h f(x)|^p d\tau(x) &= n\int_{r_{k-2}}^{r_{k+2}} \rho^{n-1}(1-\rho^2)^{-n} M_p^p(|\nabla^h f|,\rho)d\rho \\ &\le C_n C^p \int_{r_{k-2}}^{r_{k+2}} (1-\rho)^{-n-\beta p} d\rho \\ &\le C^p C_{n,\beta,p}\left(2^{(n-1)+\beta p}\right)^k.\end{aligned}$$

Hence with $K = C_{n,\beta,p}$,

$$\begin{aligned}|f(r\zeta)| &\le |f(0)| + CK\sum_{k=1}^{N}\left(2^{\frac{(n-1)}{p}+\beta}\right)^k \\ &\le |f(0)| + CK\left(2^N\right)^{\frac{(n-1)}{p}+\beta} \\ &\le \frac{CK'}{(1-r)^{\frac{(n-1)}{p}+\beta}},\end{aligned}$$

for some constant K'. Taking the supremum over $\zeta \in \mathbb{S}$ proves the result for $q = \infty$.

Suppose $0 < p < q < \infty$. Then

$$\begin{aligned}M_q(f,r) &= \left\{\int_{\mathbb{S}} |f(r\zeta)|^p |f(r\zeta)|^{q-p} d\sigma(\zeta)\right\}^{1/q} \\ &\le M_\infty(f,r)^{1-\frac{p}{q}} M_p(f,r)^{\frac{p}{q}} \\ &\le \frac{KC}{(1-r)^\lambda},\end{aligned}$$

where $\lambda = \beta + (n-1)\left(\frac{1}{p} - \frac{1}{q}\right)$. □

Our final theorem of this section is an application of the previous theorem. The proof uses the Marcinkiewicz interpolation theorem (see [75, Appendix B.1] or [99, (II) p.111]), and the method we use is similar to that used by T. M. Flett in [23, Theorem 1].

Theorem 10.6.3 *If* $1 < p < q \leq \infty$, $\alpha = \frac{1}{p} - \frac{1}{q}$, *and* $p \leq \lambda < \infty$, *then for all* $\hat{f} \in L^p(\mathbb{S})$ *there exists a constant* $C_{p,q}$, *independent of* $\hat{f}$, *such that*

$$\left[\int_0^1 (1-r)^{(n-1)\lambda\alpha-1} M_q^\lambda(f,r)dr\right]^{1/\lambda} \leq C_{p,q}\|\hat{f}\|_p,$$

where $f(x) = P_h[\hat{f}](x)$.

Proof. Define $\omega(r) = r^{n-1}$. Also, define T on $L^p(\mathbb{S})$ by

$$T\hat{f}(r) = \omega(r)^{-\frac{1}{q}} M_q(f, 1-r) \quad 0 < r < 1, \hat{f} \in L^p,$$

where $f(x) = P_h[\hat{f}](x)$. By Minkowski's inequality T is sub-additive, that is, $T(\hat{f}_1+\hat{f}_2)(r) \leq T\hat{f}_1(r)+T\hat{f}_2(r)$ for all $r \in (0,1)$. Furthermore, for $1 \leq p \leq q$, by the previous theorem,

$$M_q(f, 1-r) \leq C\|\hat{f}\|_p \omega(r)^{\frac{1}{q}-\frac{1}{p}}.$$

Therefore,

$$T\hat{f}(r) \leq C\omega(r)^{-\frac{1}{p}}\|\hat{f}\|_p.$$

Hence,

$$\{r : T\hat{f}(r) > \alpha\} \subset \left\{r : \omega(r) < \frac{C^p\|\hat{f}\|_p^p}{\alpha}\right\}.$$

Let Ω denote the inverse function of ω. Then

$$\{r : T\hat{f}(r) > \alpha\} \subset (0, b),$$

where

$$b = \min\left\{1, \Omega(C^p\|\hat{f}\|_p^p\alpha^{-p})\right\}.$$

Define the measure ν by $d\nu = \omega'(r)dr$. Then

$$\nu(\{r : T\hat{f}(r) > \alpha\}) \leq \int_0^b \omega'(t)dt = \omega(b) \leq \frac{C^p\|\hat{f}\|_p^p}{\alpha^p}.$$

Thus T is weak-type (p,p). Hence by the Marcinkiewicz interpolation theorem, T is of strong type (p,p) for $1 < p < q$, that is,

$$\left[\int_0^1 (T\hat{f})^p d\nu\right]^{1/p} = \left[\int_0^1 \omega(r)^{-\frac{p}{q}}\omega'(r)M_q^p(f, 1-r)dr\right]^{1/p},$$

which by a change of variable

$$= c_n \left[\int_0^1 (1-r)^{(1-\frac{p}{q})(n-1)-1} M_q^p(f,r)dr \right]^{1/p}$$
$$\leq A(p,q)\|\hat{f}\|_p.$$

Finally, suppose $p < \lambda < \infty$. Then

$$\int_0^1 \omega(r)^{\lambda\alpha-1}\omega'(r)M_q^\lambda(f,1-r)dr$$
$$\leq \sup_{r\in(0,1)} \{\omega(r)^\alpha M_q(f,1-r)\}^{\lambda-p} \left\{ \int_0^1 \omega(r)^{-\frac{p}{q}}\omega'(r)M_q^p(f,1-r)dr \right\}$$
$$\leq A(p,q)\|\hat{f}\|_p^{\lambda-p+p} = A(p,q)\|\hat{f}\|_p^\lambda.$$

The result now follows by the change of variable $r = (1-\rho)$. □

As an immediate consequence of the above we have the following.

Corollary 10.6.4 *Let $f \in \mathcal{H}^p$, $1 < p < \infty$. Then*

$$\left\{ \int_{\mathbb{B}} |f(x)|^{\frac{pn}{(n-1)}} d\nu(x) \right\}^{\frac{(n-1)}{pn}} \leq A_p \|f\|_p,$$

that is, $\mathcal{H}^p \subset \mathcal{B}^{\frac{pn}{(n-1)}}$ for all $p > 1$.

Proof. Take $\lambda = q = pn/(n-1)$ in the previous theorem. □

Along the same lines, taking $\lambda = q$ gives

Corollary 10.6.5 *If $f \in \mathcal{H}^p$, $1 < p < \infty$, then for all $q > p$,*

$$\left\{ \int_{\mathbb{B}} (1-|x|^2)^{(n-1)\frac{p}{q}} |f(x)|^q d\tau(x) \right\}^{\frac{1}{q}} \leq A_{p,q}\|f\|_p,$$

that is, $\mathcal{H}^p \subset \mathcal{B}^q_{(n-1)\frac{q}{p}}$ for all $q > p > 1$.

Example 10.6.6 We now use Theorem 8.6.3 to show that for $0 < p < \infty$, $\gamma > (n-1)$, there exists $h \in \mathcal{B}^p_\gamma$ which is not in $\mathcal{H}^q$ for any $q \geq 1$. Since $\gamma - n > -1$, we can choose $\alpha > 0$ such that $\gamma - n - \alpha p > -1$. Take $\omega(r) = (1-r^2)^{-\alpha}$. Then by Theorem 8.6.3 there exists an $\mathcal{H}$-harmonic function on $\mathbb{B}$ such that

(i) $|h(r\zeta)| \leq \omega(r)$ for all $r \in [0,1)$ and all $\zeta \in \mathbb{S}$, and

(ii) $h(r\zeta)$ fails to have a finite limit as $r \to 1$ at every $\zeta \in \mathbb{S}$.

As a consequence of (i) we have that $h \in \mathcal{B}^p_\gamma$, but by (ii) h is not in $\mathcal{H}^q$ for any $q \geq 1$.

10.7 Littlewood–Paley Inequalities

The classical Littlewood–Paley inequalities for harmonic functions [51] in $\mathbb{D}$ are as follows. Let h be harmonic on $\mathbb{D}$. Then there exist positive constants C_1, C_2, independent of h, such that

(a) for $1 < p \le 2$,

$$\|h\|_p^p \le C_1 \left[|h(0)|^p + \iint_{\mathbb{D}} (1-|z|)^{p-1} |\nabla h(z)|^p dx\,dy \right]. \tag{10.7.1}$$

(b) For $p \ge 2$, if $h \in \mathcal{H}^p$, then

$$\iint_{\mathbb{D}} (1-|z|)^{p-1} |\nabla h(z)|^p dx\,dy \le C_2 \|h\|_p^p. \tag{10.7.2}$$

In 1956 T. M. Flett [22] proved that for analytic functions inequality (10.7.1) is valid for all p, $0 < p \le 2$. Hence if $u = \operatorname{Re} h$, h analytic, then since $|\nabla u| = |h'|$ it immediately follows that inequality (10.7.1) also holds for harmonic functions in $\mathbb{D}$ for all p, $0 < p \le 2$. A new proof of the Littlewood–Paley inequalities for analytic functions was given by D. H. Luecking in [53]. Also, a short proof of the inequalities for harmonic functions in $\mathbb{D}$ valid for all p, $0 < p < \infty$, has been given recently by M. Pavlović in [65].

The Littlewood–Paley inequalities are also known to be valid for harmonic functions in the unit ball in $\mathbb{R}^n$ when $p > 1$. In [77] S. Stević proved that for $n \ge 3$, inequality (10.7.1) is valid for all $p \in [\frac{n-2}{n-1}, 1]$. In [87], the author proved that the analogue of (10.7.1) was valid for harmonic functions on bounded domains with $C^{1,1}$ boundary for all p, $0 < p \le 2$. In the case of the unit ball, this result was subsequently improved upon by O. Djordjević and M. Pavlović. In [17] they proved that if u is a harmonic function on the unit ball $\mathbb{B}$ in $\mathbb{R}^N$ and if $0 < p \le 1$, then

$$\int_{\mathbb{S}} u^*(y)^p d\sigma(y) \le C \left(|u(0)|^p + \int_{\mathbb{B}} (1-|x|)^{p-1} |\nabla^h u(x)|^p dx \right), \tag{10.7.3}$$

where u^* is the non-tangential maximal function of u. The Littlewood–Paley inequalities have also been extended by the author to Hardy–Orlicz spaces of harmonic functions on general domains in $\mathbb{R}^n$, including Lipschitz domains [88], as well as $\mathcal{H}$-harmonic functions on the unit ball $\mathbb{B}$ [89].

In the present section we consider the Littlewood–Paley inequalities for non-negative $\mathcal{H}$-subharmonic functions on $\mathbb{B}$ for which $\Delta_h f$ is quasi-nearly $\mathcal{H}$-subharmonic on $\mathbb{B}$. The results will then apply to the special case where $f = |h|^2$, where h is $\mathcal{H}$-harmonic on $\mathbb{B}$, or more generally to $f = \sum |h_n|^2$, with suitable convergence, where each h_n is again $\mathcal{H}$-harmonic on $\mathbb{B}$. In each case,

$\Delta_h f$ is quasi-nearly $\mathcal{H}$-subharmonic. Our main results of the section are the following theorem and corollaries.

Theorem 10.7.1 *Let f be a non-negative C^2 $\mathcal{H}$-subharmonic function on $\mathbb{B}$ for which $\Delta_h f$ is quasi-nearly $\mathcal{H}$-subharmonic on $\mathbb{B}$. Then there exist constants C_1 and C_2, independent of f, such that*

(a) *for $1 \le p < \infty$,*

$$\int_{\mathbb{B}} (1-|x|^2)^{n-1}(\Delta_h f(x))^p d\tau(x) \le C_1 \|f\|_p.$$

(b) *For $0 < p \le 1$, if in addition f^p is C^2 and $\mathcal{H}$-subharmonic, then*

$$\|f\|_p^p \le C_2\left[f^p(0) + \int_{\mathbb{B}} (1-|x|^2)^{n-1}(\Delta_h f(x))^p d\tau(x)\right].$$

Corollary 10.7.2 *Let h be $\mathcal{H}$-harmonic on $\mathbb{B}$. Then there exist constants C_1 and C_2, independent of h, such that*

(a) *if $f \in \mathcal{H}^p$, $2 \le p < \infty$, then*

$$\int_{\mathbb{B}} (1-|x|^2)^{n-1}|\nabla^h h(x)|^p d\tau(x) \le C_1 \|h\|_p^p.$$

(b) *For $0 < p \le 2$, and all $\alpha > 1$,*

$$\int_{\mathbb{S}} (M_\alpha h)^p d\sigma \le C_2\left[|h(0)|^p + \int_{\mathbb{B}} (1-|x|^2)^{n-1}|\nabla^h h(x)|^p d\tau(x)\right].$$

The previous corollary can also be stated as follows:

Corollary 10.7.3 *Let h be $\mathcal{H}$-harmonic on $\mathbb{B}$.*

(a) *If $h \in \mathcal{H}^p$, $2 \le p < \infty$, then $h \in \mathcal{D}_{n-1}^p$ with $\|h\|_{\mathcal{D}_{n-1}^p} \le C\|h\|_p$.*

(b) *If $h \in \mathcal{D}_{n-1}^p$, $0 < p \le 2$, then $h \in \mathcal{H}^p$ with $\|M_\alpha h\|_p \le C\|h\|_{\mathcal{D}_{n-1}^p}$.*

As a consequence of the previous corollary, a function $h \in \mathcal{H}^2$ if and only

$$\int_{\mathbb{B}} (1-|x|^2)^{n-1}|\nabla^h h(x)|^2 d\tau(x)$$

is finite. A similar result follows from Exercise 10.8.9. If for $\alpha > 1$ we set

$$S_\alpha(h,\zeta) = \left(\int_{\Gamma_\alpha(\zeta)} |\nabla^h h(x)|^2 d\tau(x)\right)^{\frac{1}{2}},$$

then it is easily shown that

$$\int_{\mathbb{S}} S_\alpha^2(h,\zeta)d\sigma(\zeta) \approx \int_{\mathbb{B}} (1-|x|^2)^{n-1}|\nabla^h h(x)|^2 d\tau(x).$$

Thus $h \in \mathcal{H}^2$ if and only if $S_\alpha \in L^2(\mathbb{S})$. The function $S_\alpha(h, \zeta)$ is called the **square area integral** of h. This characterization of $\mathcal{H}^2$ was also proved by P. Cifuentis in [15]. For related problems the reader is referred to Exercise 10.8.10.

For the proof of the main theorem we require several preliminary lemmas.

Lemma 10.7.4 *Let f be a positive C^2 $\mathcal{H}$-subharmonic function on $\overline{B(0,\rho)}$, $0 < \rho < \frac{1}{2}$, such that $\Delta_h f$ is quasi-nearly $\mathcal{H}$-subharmonic on $B(0,\rho)$. Then*
(a) *for $p \geq 1$,*

$$\int_{B_{\rho/2}} (\Delta_h f(y))^p d\tau(y) \leq c_n \rho^{2-2p} \int_{B_\rho} \Delta_h f^p(y) d\tau(y).$$

(b) *Suppose $0 < p \leq 1$. If in addition, f^p is C^2 and $\mathcal{H}$-subharmonic, then*

$$\int_{B_{\rho/4}} \Delta_h f^p(y) d\tau(y) \leq c_n \rho^{2p-2} \int_{B_\rho} (\Delta_h f(y))^p d\tau(y).$$

Proof. Without loss of generality we assume that $n \geq 3$. The case $n = 2$ is proved similarly. Recall from Theorem 4.1.1, for any r, $0 < r < 1$, if f^p is $\mathcal{H}$-subharmonic and C^2, then

$$\int_{\mathbb{S}} f^p(r\zeta) d\sigma(\zeta) - f^p(0) = \int_{B_r} g(|x|, r) \Delta_h f^p(x) d\tau(x), \tag{10.7.4}$$

where

$$g(|x|, r) = \frac{1}{n} \int_{|x|}^{r} \frac{(1 - s^2)^{n-2}}{s^{n-1}} ds.$$

We first note that for $|x| \leq \frac{\delta}{2}$,

$$g(|x|, \delta) \geq c_{n,\delta} \delta^{2-n}. \tag{10.7.5}$$

(a) **The case** $p \geq 1$. Set $\delta = \frac{1}{2}\rho$. Since $\Delta_h f$ is quasi-nearly $\mathcal{H}$-subharmonic,

$$(\Delta_h f(0))^p \leq c_n \left(\rho^{-n} \int_{B_{\rho/2}} \Delta_h f(x) d\tau(x) \right)^p,$$

which by inequality (10.7.5) and identity (10.7.4)

$$\begin{aligned} &\leq c_{n,\delta} \left(\rho^{-2} \int_{B_\delta} g(|x|, \delta) \Delta_h f(x) d\tau(x) \right)^p \\ &= c_{n,\delta} \rho^{-2p} \left[\int_{\mathbb{S}} f(\delta\zeta) d\sigma(\zeta) - f(0) \right]^p. \end{aligned}$$

Since $(a-b)^p \le a^p - b^p$ whenever $p \ge 1$ and $0 \le b \le a$,

$$\begin{aligned}(\Delta_h f(0))^p &\le c_{n,\delta}\rho^{-2p}\left[\left(\int_{\mathbb{S}} f(\delta\zeta)d\sigma(\zeta)\right)^p - f^p(0)\right] \\ &\le c_{n,\delta}\left[\int_{\mathbb{S}} f^p(\delta\zeta) - f^p(0)\right].\end{aligned}$$

The last inequality follows by Hölder's inequality. Hence by (10.7.4) we have

$$(\Delta_h f(0))^p \le c_{n,\delta}\rho^{-2p}\int_{B_\delta} g(|x|,\delta)\Delta_h f^p(x)d\tau(x).$$

Replacing f by $f\circ\varphi_y$, $y \in B_\delta$, gives

$$(\Delta_h f(y))^p \le c_{n,\delta}\rho^{-2p}\int_{B_\delta} g(|x|,\delta)\Delta_h(f(\varphi_y(x)))^p d\tau(x),$$

which by the change of variable $w = \varphi_y(x)$

$$= c_{n,\delta}\rho^{-2p}\int_{E_\delta(y)} g(|\varphi_y(w)|,\delta)\Delta_h f^p(w)d\tau(w).$$

Since $E_\delta(y) \subset B_\rho$ for all $y \in B_\delta$,

$$\int_{B_\delta}(\Delta_h f(y))^p d\tau(y) \le c_{n,\delta}\rho^{-2p}\int_{B_\delta}\int_{B_\rho} g(|\varphi_y(w)|,\delta)\Delta_h f^p(w)d\tau(w)d\tau(y),$$

which by Fubini's theorem

$$= c_{n,\delta}\rho^{-2p}\int_{B_\rho}\Delta_h f^p(w)\int_{B_\delta} g(|\varphi_y(w)|,\delta)d\tau(y)d\tau(w).$$

But $|\varphi_y(w)| = |\varphi_w(y)|$. Therefore since $E(w,\delta) \subset B_{2\rho}$,

$$\int_{B_\delta} g(|\varphi_w(y)|,\delta)\,d\tau(y) = \int_{E_\delta(w)} g(|x|,\delta)\,d\tau(x) \le \int_{B_{2\rho}} g(|x|,\delta)\,d\tau(x) \le C_n\rho^2.$$

Therefore,

$$\int_{B_{\rho/2}}(\Delta_h f(y))^p d\tau(y) \le c_{n,\delta}\rho^{2-2p}\int_{B_\rho}\Delta_h f^p(w)d\tau(w),$$

which proves (a).

(b) **The case** $0 < p \le 1$**.** For this case we have to assume in addition that f^p is C^2 and $\mathcal{H}$-subharmonic. As above, set $\delta = \frac{1}{2}\rho$. Then by inequality (10.7.5),

$$\int_{B_{\delta/2}} f^p(y)d\tau(y) \le c_{n,\delta}\delta^{n-2}\int_{B_\delta} g(|x|,\delta)\Delta_h f^p(y)d\tau(y),$$

which by identity (10.7.4)

$$= c_{n,\delta}\delta^{n-2}\left[\int_{\mathbb{S}} f^p(\delta\zeta)d\sigma(\zeta) - f^p(0)\right].$$

Since $0 < p < 1$,

$$\begin{aligned}
\left[\int_{\mathbb{S}} f^p(\delta\zeta)d\sigma(\zeta) - f^p(0)\right] &\le \left[\int_{\mathbb{S}} f(\delta\zeta)d\sigma(\zeta) - f(0)\right]^p \\
&= \left[\int_{B_\delta} g(|x|,\delta)\Delta_h f(x)d\tau(x)\right]^p \\
&\le \sup_{x\in B_\delta}(\Delta_h f(x))^p\left(\int_{B_\delta} g(|x|,\delta)d\tau(x)\right)^p \\
&\le C_n\rho^{2p}\sup_{x\in B_\delta}(\Delta_h f(x))^p.
\end{aligned}$$

Since $\Delta_h f$ is quasi-nearly $\mathcal{H}$-subharmonic and $E_\delta(x) \subset B_\rho$ for all $x \in B_\delta$,

$$\rho^{2p}(\Delta_h f(x))^p \le C\rho^{2p-n}\int_{B_\rho}(\Delta_h f(x))^p d\tau(x).$$

Therefore,

$$\int_{B_{\rho/4}} \Delta_h f^p(y)d\tau(y) \le C_{n,\delta}\rho^{2p-2}\int_{B_\rho}(\Delta_h f(x))^p d\tau(x).$$

□

Lemma 10.7.5 *Let f be a positive C^2 $\mathcal{H}$-subharmonic function on $\mathbb{B}$ such that $\Delta_h f$ is quasi-nearly $\mathcal{H}$-subharmonic on $\mathbb{B}$. Fix δ, $0 < \delta < \frac{1}{2}$, and $\gamma \in \mathbb{R}$. Then*

(a) *for $p \ge 1$,*

$$\int_{\mathbb{B}}(1-|x|^2)^\gamma(\Delta_h f(x))^p d\tau(x) \le C_{n,\delta}\int_{\mathbb{B}}(1-|x|^2)^\gamma \Delta_h f^p(x)d\tau(x)$$

where $C_{n,\delta}$ is a constant independent of f.

(b) *For $0 < p < 1$, if in addition f^p is $\mathcal{H}$-subharmonic and C^2, then*

$$\int_{\mathbb{B}}(1-|x|^2)^\gamma \Delta_h f^p(x)d\tau(x) \le C_{n,\delta}\int_{\mathbb{B}}(1-|x|^2)^\gamma(\Delta_h f(x))^p d\tau(x).$$

Proof. (a) By Lemma 10.3.1

$$\begin{aligned}
&\int_{\mathbb{B}}(1-|x|^2)^\gamma(\Delta_h f(x))^p d\tau(x) \\
&\qquad \le C_n\int_{\mathbb{B}}(1-|w|^2)^\gamma\left[\int_{E_{\delta/2}(w)}(\Delta_h f(x))^p d\tau(x)\right]d\tau(w).
\end{aligned}$$

But by Lemma 10.7.4(a)

$$\begin{aligned}\int_{E_{\delta/2}(w)} (\Delta_h f(x))^p d\tau(x) &= \int_{B_{\delta/2}} (\Delta_h(f(\varphi_w(y)))^p d\tau(y) \\ &\leq C_{n,\delta} \int_{B_\delta} \Delta_h(f \circ \varphi_w)^p(y) d\tau(y) \\ &= C_{n,\delta} \int_{E_\delta(w)} \Delta_h f^p(y) d\tau(y).\end{aligned}$$

Applying Lemma 10.3.1 again gives

$$\int_{\mathbb{B}} (1-|w|^2)^\gamma \left[\int_{E_\delta(w)} \Delta_h f^p(y) d\tau(y)\right] d\tau(w) \leq C_n \int_{\mathbb{B}} (1-|w|^2)^\gamma \Delta_h f^p(w) d\tau(w).$$

The proof of (b) follows similarly. □

Proof of Theorem 10.7.1 (a) **The case** $p \geq 1$. Assume first that $f(x) > 0$ for all $x \in \mathbb{B}$. Then f^p is C^2 for all $p > 0$. Fix δ, $0 < \delta < \frac{1}{2}$. Then with $\gamma = n - 1$,

$$\int_{\mathbb{B}} (1 - |x|^2)^{n-1} (\Delta_h f(x))^p d\tau(x) \leq C_{n,\delta} \int_{\mathbb{B}} (1 - |x|^2)^{n-1} \Delta_h f^p(x) d\tau(x),$$

which by Theorem 4.1.1

$$\leq C_{n,\delta} \lim_{r\to 1} \int_{\mathbb{S}} f^p(rt) d\sigma(t) = C_{n,\delta} \|f\|_p^p.$$

For the general case $f(x) \geq 0$ take $f_\epsilon(x) = f(x) + \epsilon$. Then $\Delta_h f_\epsilon = \Delta_h f$, and by the above

$$\int_{\mathbb{B}} (1 - |x|^2)^{n-1} (\Delta_h f(x))^p d\tau(x) \leq C_{n,\delta} \|f_\epsilon\|_p^p.$$

Letting $\epsilon \to 0$ proves the result.

(b) **The case** $0 < p \leq 1$. Assume that $\int_{\mathbb{B}} (1-|x|^2)^{n-1} (\Delta_h f(x))^p d\tau(x) < \infty$. Then by Lemmas 10.3.2 and 10.7.5,

$$\begin{aligned}\int_{\mathbb{S}} f^p(rt) d\sigma(t) &\leq f^p(0) + \frac{1}{n} \int_{\mathbb{B}} \frac{(1 - |x|^2)^{n-1}}{|x|^{n-2}} \Delta_h f^p(x) d\tau(x) \\ &\leq f^p(0) + \sup_{x \in B_{\frac{1}{4}}} \Delta_h f^p(x) + C_n \int_{\mathbb{B}} (1 - |x|^2)^{n-1} (\Delta_h f(x))^p d\tau(x).\end{aligned}$$

Hence $f \in \mathcal{S}^p$ (or alternately $f^p \in \mathcal{S}^1$) and thus has an $\mathcal{H}$-harmonic majorant H_{f^p} on $\mathbb{B}$ with $\|f\|_p^p = H_{f^p}(0)$. Hence by the Riesz decomposition theorem,

$$H_{f^p}(x) = f^p(x) + \int_{\mathbb{B}} G_h(x, y) \Delta_h f^p(y) d\tau(y).$$

Thus by Remark 9.2.5(c),

$$\|f\|_p^p \leq f^p(0) + C_n \int_{\mathbb{B}} (1 - |y|^2)^{n-1} \Delta_h f^p(y) d\tau(y),$$

which by Lemma 10.7.5

$$\leq f^p(0) + C_{n,\delta} \int_{\mathbb{B}} (1 - |y|^2)^{n-1} (\Delta_h f(y))^p d\tau(y).$$

Therefore

$$\|f\|_p^p \leq f^p(0) + C_{n,\delta} \int_{\mathbb{B}} (1 - |y|^2)^{n-1} (\Delta_h f(y))^p d\tau(y),$$

which proves the result. □

Proof of Corollary 10.7.2 For (a) take $f = |h|^2$. Then $\Delta_h f = 2|\nabla^h h|^2$. For $h \in \mathcal{H}^q$, $q \geq 2$, the result follows from Theorem 10.7.1(a) by taking $p = q/2$.

(b) For $0 < q \leq 1$ the result is Theorem 8.5.1. For $1 < q < 2$ we consider $f_\epsilon = |h + i\epsilon|^2$, $\epsilon > 0$. Then f_ϵ is a positive C^2 $\mathcal{H}$-subharmonic function for which $\Delta_h f_\epsilon = 2|\nabla^h f|^2$. Thus by Theorem 10.7.1(b) with $p = q/2$,

$$\|f_\epsilon\|_p^p \leq C\left[|h(0) + i\epsilon|^q + \int_{\mathbb{B}} (1 - |x|^2)^{n-1} |\nabla^h h(x)|^q d\tau(x)\right].$$

The result now follows by letting $\epsilon \to 0$ and using the fact that for $q > 1$, $\int_{\mathbb{S}} (M_\alpha h)^q d\sigma \leq A_q \int_{\mathbb{S}} |h|^q d\sigma$. □

10.8 Exercises

10.8.1. If $f \in \mathcal{H}^p$, $p \geq 1$ and $p < q \leq \infty$, prove that

$$M_q(f, r) = o\left((1 - r)^{(n-1)(\frac{1}{p} - \frac{1}{q})}\right).$$

10.8.2. Fix $\beta > 1$. Show that there exists a non-negative continuous $\mathcal{H}$-superharmonic function V on $\mathbb{B}$ such that $V(x) > 0$ for all $x \in \Gamma_\beta(e_1)$ and $V(x) = 0$ on $\mathbb{B} \setminus \Gamma_\beta(e_1)$.

10.8.3. (*) Fix $\beta > 1$. **Question.** Does there exist a positive $\mathcal{H}$-harmonic function h on $\Gamma_\beta(e_1)$ such that

(a) $h(x) \leq C|x - e_1|^{1-n-\epsilon(\beta)}$, $\epsilon(\beta) > 0$, where $\epsilon(\beta) \to 0$ as $\beta \to 1$, and

(b) $h(x) = 0$ on $\partial\Gamma_\beta(e_1)$?

If the answer to the previous exercise is yes, then it can be used to provide a positive answer to the following question.

10.8.4. (*) For $0 < p < \frac{1}{2}$ and $\gamma > p(n-1)$, does there exist a non-negative $\mathcal{H}$-subharmonic function f with $f \in L^p_\gamma(\tau)$?

10.8.5. (a) Suppose f is a non-negative C^2 $\mathcal{H}$-subharmonic function such that f^p is also C^2 for all $p > 1$. Prove that

$$\|f\|^p_{\gamma,p} \approx |f(0)|^p + \int_{\mathbb{B}} (1-|x|^2)^\gamma \Delta_h f^p(x) d\tau(x).$$

(b) Let h be an $\mathcal{H}$-harmonic function on $\mathbb{B}$. Prove that for $p > 1$, $h \in \mathcal{B}^p_\gamma$ if and only if $\int_{\mathbb{B}} (1-|x|^2)^\gamma |h(x)|^{p-2} |\nabla^h h(x)|^2 d\tau(x) < \infty$, and if this is the case, then

$$\|h\|^p_{\gamma,p} \approx |h(0)|^p + \int_{\mathbb{B}} (1-|x|^2)^\gamma |h(x)|^{p-2} |\nabla^h h(x)|^2 d\tau(x).$$

10.8.6. As in [36] investigate properties of the operators B_α.

10.8.7. (*) Using the methods of [64], prove Theorem 10.7.1 using the Riesz measure μ_f rather than $\Delta_h f$.

10.8.8. (*) Using the methods of [88], prove the following version of the **Littlewood–Paley inequalities for Hardy-Orlicz spaces** of $\mathcal{H}$-harmonic functions.

Theorem *Let $\psi \geq 0$ be an increasing convex C^2 function on $[0, \infty)$ satisfying $\psi(0) = 0$ and $\psi(2x) \leq c\psi(x)$ for some positive constant c. Set $\varphi(t) = \psi(\sqrt{t})$. Then there exist constants C_1 and C_2 such that for all $\mathcal{H}$-harmonic functions f,*

(a) *if φ is concave on $[0, \infty)$,*

$$\|f\|_\psi \leq C_1 \left[\psi(|f(0)|) + \int_{\mathbb{B}} (1-|x|^2)^{n-1} \psi(|\nabla^h f(x)|) d\tau(x) \right].$$

(b) *If φ is convex on $[0, \infty)$, then*

$$C_2 \left[\psi(|f(0)|) + \int_{\mathbb{B}} (1-|x|^2)^{n-1} \psi(|\nabla^h f(x)|) d\tau(x) \right] \leq \|f\|_\psi.$$

In the above, $\|f\|_\psi = \lim_{r \to 1} \int_{\mathbb{S}} \psi(|f(rt)|) d\sigma(t)$.

10.8.9. **Hardy and Dirichlet spaces.** [89] For $\gamma \in \mathbb{R}$ the **Dirichlet space** $\mathcal{D}_\gamma$ is defined as the set of C^1 functions on $\mathbb{B}$ for which $|\nabla^h f| \in L^2_\gamma(\tau)$ with norm

$$\|f\|_{\mathcal{D}_\gamma} = |f(0)| + D_\gamma(f)^{1/2}, \qquad (10.8.1)$$

where $D_\gamma(f)$ is given by

$$D_\gamma(f) = \int_{\mathbb{B}} (1-|x|^2)^\gamma |\nabla^h f(x)|^2 d\tau(x).$$

Prove the following.

Theorem *Let f be $\mathcal{H}$-harmonic on $\mathbb{B}$.*
(a) *If $f \in \mathcal{D}_\gamma$ for some γ, $(n-3) < \gamma \le (n-1)$, $(0 < \gamma \le 1$ when $n = 2)$, then $f \in \mathcal{H}^p$ for $p = 2(n-1)/\gamma$, with*

$$\|f\|_p \le C_1 \|f\|_{\mathcal{D}_\gamma},$$

where C_1 is a positive constant independent of f.
(b) *If $f \in \mathcal{H}^p$ for some p, $1 < p \le 2$, then $f \in \mathcal{D}_\gamma$ for $\gamma = 2(n-1)/p$ with*

$$\|f\|_{\mathcal{D}_\gamma} \le C_2 \|f\|_p,$$

where C_2 is a positive constant independent of f.

10.8.10. **Littlewood–Paley theory for $\mathcal{H}$-subharmonic functions.** In 1930, N. Lusin [54] introduced the square area integral S_α, $\alpha > 1$, for an analytic function f on $\mathbb{D}$ given by

$$S_\alpha(f,\zeta)^2 = \int_{\Gamma_\alpha(\zeta)} |f'(z)|^2 dA(z), \qquad \zeta \in [0, 2\pi).$$

In 1938 J. Marcinkiewicz and A. Zygmund [56] proved that if f is in the Hardy space H^p, $0 < p < \infty$, then $S_\alpha \in L^p([0, 2\pi))$ with

$$\int_0^{2\pi} S_\alpha^p d\theta \le A_{\alpha,p} \int_0^{2\pi} |f|^p d\theta,$$

and that the reverse inequality holds for $p > 1$. For $\mathcal{H}$-harmonic functions h on $\mathbb{B}$, the analogue of the Lusin **square area integral** is the function $S_\alpha(h,\zeta)$ defined for $\zeta \in \mathbb{S}$ by

$$S_\alpha(h) = S_\alpha(h,\zeta) = \left(\int_{\Gamma_\alpha(\zeta)} |\nabla^h h(x)|^2 d\tau(x)\right)^{\frac{1}{2}}. \qquad (10.8.2)$$

This function was considered by S. Grellier and P. Jaming in [31]. The main result of their paper was that $S_\alpha \in L^p(\mathbb{S})$, $0 < p < \infty$, $\alpha > 1$, if and only if the non-tangential maximal function $M_\alpha h \in L^p(\mathbb{S})$, with equivalence of norms. For a non-negative C^2 $\mathcal{H}$-subharmonic function f, set

$$S_\alpha(f) = S_\alpha(f,\zeta) = \left(\int_{\Gamma_\alpha(\zeta)} \Delta_h f^2(x) d\tau(x)\right)^{\frac{1}{2}}. \qquad (10.8.3)$$

Taking $f = |h|$ where h is $\mathcal{H}$-harmonic on $\mathbb{B}$ gives, up to a constant, (10.8.2).

(a) As in [91], prove the following theorem.

Theorem *Let f be a non-negative C^2 $\mathcal{H}$-subharmonic function on $\mathbb{B}$ such that f^{p_o} is $\mathcal{H}$-subharmonic for some $p_o > 0$. If $f \in \mathcal{S}^p$ for some $p > p_o$, then for for every $\alpha > 1$,*

$$\int_{\mathbb{S}} S^p_\alpha(f,\zeta)d\sigma(\zeta) \le A_\alpha \|f\|^p_p$$

for some constant A_α independent of f.

(b) (*) Prove the reverse inequality for $\mathcal{H}$-subharmonic functions.

10.8.11. (a) For $\gamma > (n-1)$ prove that $\mathcal{B}^2_\gamma$ is a real Hilbert space with inner product

$$\langle f,g\rangle_\gamma = \int_{\mathbb{B}} (1-|x|^2)^\gamma f(x)g(x)d\tau(x). \tag{10.8.4}$$

(b) Using the fact that point evaluation is a bounded linear functional on $\mathcal{B}^2_\gamma$, prove that there exists a function $R^\gamma(x,y)$ on $\mathbb{B}\times\mathbb{B}$ satisfying the following:

i. $R^\gamma(x,y) = R^\gamma(y,x)$.

ii. For each $y \in \mathbb{B}$, prove that the function R^γ_y defined by $R^\gamma_y(x) = R^\gamma(x,y)$ is in $\mathcal{B}^2_\gamma$.

iii. $f(y) = \langle f, R^\gamma_y\rangle = \int_{\mathbb{B}} (1-|x|^2)^\gamma R^\gamma(x,y)d\tau(x)$ for all $f \in \mathcal{B}^2_\gamma$.

The function R^γ is called the **reproducing kernel** of $\mathcal{B}^2_\gamma$.

(c) (*) Find the function $R^\gamma(x,y)$. (See [10, Theorem 8.13] for the analogous question for Euclidean harmonic functions on $\mathbb{B}$.)

10.8.12. Define the **Bloch space** $\mathcal{B}$ of $\mathcal{H}$-harmonic functions on $\mathbb{B}$ by

$$\mathcal{B} = \left\{ f\ \mathcal{H}\text{-harmonic} : \sup_{x\in\mathbb{B}} |\nabla^h f(x)| < \infty \right\}$$

with norm

$$\|f\|_{\mathcal{B}} = |f(0)| + \sup_{x\in\mathbb{B}} |\nabla^h f(x)|.$$

The **little Bloch space** $\mathcal{B}_0$ is defined as

$$\mathcal{B}_0 = \left\{ f \in \mathcal{B} : \lim_{|x|\to 1} |\nabla^h f(x)| = 0 \right\}.$$

Prove the following:

(a) $\mathcal{B}$ with the seminorm $\sup_{x\in\mathbb{B}} |\nabla^h f(x)|$ is Möbius invariant.

(b) $\mathcal{B}$ is a Banach space.

(c) $\mathcal{B}_0$ is a closed subspace of $\mathcal{B}$.

10.8.13. (*) Does there exist a Möbius invariant Hilbert space H of $\mathcal{H}$-harmonic functions on $\mathbb{B}$? If the answer is yes do the following:

(a) Provide a characterization of H.
(b) Find the reproducing kernel of H.

10.8.14. (*) For $0 < \mu < (n-1)$, let K_μ be the kernel on $\mathbb{S} \times \mathbb{S}$ given by

$$K_\mu(\zeta, \eta) = \int_0^1 (1-r^2)^{\mu-1} P_h(r\zeta, \eta)dr.$$

Note: K_μ is the fractional integral of P_h. As in [29] investigate the mapping

$$K_\mu f(\zeta) = \int_{\mathbb{S}} K_\mu(\zeta, \eta) f(\eta) d\sigma(\eta).$$

It is conjectured that the mapping $f \to K_\mu f$ maps $L^p(\mathbb{S})$ to $L^q(\mathbb{S})$ where

$$\frac{1}{q} = \frac{1}{p} - \frac{\mu}{(n-1)}.$$

10.8.15. (*) A classical theorem in the theory of several complex variables states that if f is holomorphic in the unit ball $\mathbb{B}$ of $\mathbb{C}^n$ and $Rf = \sum z_j \frac{\partial f}{\partial z_j} \in H^p$ for some p, $0 < p < n$, then $f \in H^{np/(n-p)}$ (e.g., [29]). Suppose f is $\mathcal{H}$-harmonic on $\mathbb{B} \subset \mathbb{R}^n$. If $T_{i,j} f \in \mathcal{H}^p$ for some p, what can be said about f?

References

[1] M. Abramowitz and I. Stegun. *Handbook of Mathematical Functions*. Applied Math. Series 55. National Bureau of Standards, 1964.

[2] P. Ahern, J. Bruna, and C. Cascante. H^p-theory for generalized $\mathcal{M}$-harmonic functions on the unit ball. *Indiana Univ. Math. J.*, 45(1):103–135, 1996.

[3] P. Ahern, M. Flores, and W. Rudin. An invariant volume mean value property. *J. Funct. Analysis*, 111:380–397, 1993.

[4] L. Ahlfors. *Möbius Transformations in Several Dimensions*. University of Minnesota, School of Mathematics, 1981.

[5] L. V. Ahlfors. Hyperbolic motions. *Nagoya Math. J.*, 29:163–166, 1967.

[6] H. Aikawa. Tangential behavior of Green potentials and contractive properties of L^p-potentials. *Tokyo J. Math.*, 9:223–245, 1986.

[7] J. Arazy and S. Fisher. The uniqueness of the Dirichlet space among Möbius invariant function spaces. *Illinois J. Math.*, 29:449–462, 1985.

[8] D. H. Armitage. On the global integrability of superharmonic functions. *J. London Math. Soc.*, 4:365–373, 1971.

[9] M. Arsove and A. Huber. On the existence of non-tangential limits of subharmonic functions. *J. London Math. Soc.*, 42:125–132, 1967.

[10] S. Axler, P. Bourdon, and W. Ramey. *Harmonic Function Theory*. Springer-Verlag, New York, NY, 1992.

[11] A. F. Beardon. *The Geometry of Discrete Groups*. Springer-Verlag, New York, NY, 1983.

[12] A. P. Calderón. Commutators of singular integral operators. *Proc. Nat. Acad. Sci. U. S. A.*, 53:1092–1099, 1965.

[13] I. Chavel. *Eigenvalues in Riemannian Geometry*. Academic Press, Orlando, FL, 1984.

[14] P. Cifuentis. H^p classes on rank one symmetric spaces of noncompact type. II. Nontangential maximal function and area integral. *Bull. Sci. Math.*, 108:355–371, 1984.

[15] P. Cifuentis. A characterization of H^2 classes on rank one symmetric spaces of noncompact type. *Proc. Amer. Math. Soc.*, 106:519–525, 1989.

[16] J. A. Cima and C. S. Stanton. Admissible limits of $\mathcal{M}$-subharmonic functions. *Michigan Math. J.*, 32:211–220, 1985.

[17] O. Djordjević and M. Pavlović. On a Littlewood–Paley type inequality. *Proc. Amer. Math. Soc.*, 135:3607–3611, 2007.

[18] N. Dunford and J. T. Schwartz. *Linear Operators Part I*. Interscience Publishers, Inc., New York, NY, 1957.

[19] P. Duren. *Theory of H^p Spaces*. Academic Press, New York, NY, 1970.

[20] A. Erdélyi, editor. *Higher Transcendental Functions, Bateman Manuscript Project*, volume I. McGraw-Hill, New York, NY, 1953.

[21] C. Fefferman and E. Stein. H^p spaces of several variables. *Acta Math*, 129: 137–193, 1972.

[22] T. M. Flett. On some theorems of Littlewood and Paley. *J. London Math. Soc.*, 31:336–344, 1956.

[23] T. M. Flett. On the rate of growth of mean values of holomorphic functions. *Proc. London Math. Soc.*, 20:749–768, 1970.

[24] H. Furstenberg. A Poisson formula for semisimple Lie groups. *Ann. Math.*, 77:335–386, 1963.

[25] S. J. Gardiner. Growth properties of potentials in the unit ball. *Proc. Amer. Math. Soc.*, 103:861–869, 1988.

[26] L. Garding and L. Hörmander. Strongly subharmonic functions. *Math. Scand.*, 15:93–96, 1964.

[27] J. B. Garnett. *Bounded Analytic Functions*. Pure and Applied Mathematics. Academic Press, New York, NY, 1981.

[28] F. W. Gehring. On the radial order of subharmonic functions. *J. Math. Soc. Japan*, 9:77–79, 1957.

[29] I. Graham. The radial derviative, fractional integrals, and the comparitive growth of means of holomorphic functions on the unit ball in $\mathbb{C}^n$. *Annals Math. Studies*, 100:171–178, 1981.

[30] M. D. Greenberg. *Ordinary Differential Equations*. Wiley, Hoboken, NJ, 2014.

[31] S. Grellier and P. Jaming. Harmonic functions on the real hyperbolic ball II. Hardy–Sobolev and Lipschitz spaces. *Math. Nachr.*, 268:50–73, 2004.

[32] K. T. Hahn and D. Singman. Boundary behavior of invariant Green's potentials on the unit ball of $\mathbb{C}^n$. *Trans. Amer. Math. Soc.*, 309:339–354, 1988.

[33] D. J. Hallenbeck. Radial growth of subharmonic functions. *Pitman Research Notes*, 262:113–121, 1992.

[34] G. H. Hardy and J. E. Littlewood. Some properties of fractional integrals, II. *Math. Z.*, 34:403–439, 1932.

[35] J. H. Hardy and J. E. Littlewood. The strong summability of Fourier series. *Fund. Math.*, 25:162–189, 1935.

[36] H. Hedenmalm, B. Korenblum, and K. Zhu. *Theory of Bergman Spaces*, volume 199 of Graduate Texts in Mathematics. Springer, New York, NY, 2000.

[37] M. Heins. The minimum modulus of a bounded analytic function. *Duke Math. J.*, 14:179–215, 1947.

[38] S. Helgason. *Groups and Geometric Analysis*. American Mathematical Society, Providence, RI, 2000.

[39] E. Hewitt and K. Stromberg. *Real and Abstract Analysis*. Springer-Verlag, New York, NY, 1965.

[40] L. Hörmander. *Linear Partial Differential Operators*. Springer-Verlag, New York, NY, 1963.

[41] P. Jaming. *Trois problémes d'analyse harmonique*. PhD thesis, Université d'Orléans, 1998.

[42] P. Jaming. Harmonic functions on the real hyperbolic ball I. Boundary values and atomic decomposition of Hardy spaces. *Colloq. Math.*, 80:63–82, 1999.

[43] P. Jaming. Harmonic functions on classical rank one balls. *Boll. Unione Mat. Italia*, 8:685–702, 2001.

[44] M. Jevtić. Tangential characterizations of Hardy and mixed-norm spaces of harmonic functions on the real hyperbolic ball. *Acta Math. Hungar.*, 113: 119–131, 2006.

[45] A. W. Knapp. Fatou's theorem for symmetric spaces, I. *Ann. Math*, 88(2): 106–127, 1968.

[46] A. Koranyi. Harmonic functions on Hermitian hyperbolic space. *Trans. Amer. Math. Soc.*, 135:507–516, 1969.

[47] A. Koranyi. Harmonic functions on symmetric spaces. In W. M. Boothby and G. L. Weiss, editors, *Symmetric Spaces*, Marcel Dekker, Inc., New York, NY, 1972.

[48] A. Koranyi and R. P. Putz. Local Fatou theorem and area theorem for symmetric spaces of rank one. *Trans. Amer. Math. Soc.*, 224:157–168, 1976.

[49] Ü. Kuran. Subharmonic behavior of $|h|^p$ $(p > 0, h$ harmonic). *J. London Math. Soc.*, 8:529–538, 1974.

[50] N. N. Lebedev. *Special Functions and their Applications*. Dover Publications, New York, NY, 1972.

[51] J. E. Littlewood and R. E. A. C. Paley. Theorems on Fourier series and power series. *Proc. London Math. Soc.*, 42:52–89, 1936.

[52] D. H. Luecking. Boundary behavior of Green potentials. *Proc. Amer. Math. Soc.*, 96:481–488, 1986.

[53] D. H. Luecking. A new proof of an inequality of Littlewood and Paley. *Proc. Amer. Math. Soc.*, 103(3):887–893, 1988.

[54] N. Lusin. Sur une propriété des fonctions à carré sommable. *Bull. Calcutta Math. Soc.*, 20:139–154, 1930.

[55] B. D. MacCluer. Compact composition operators of $H^p(B_n)$. *Michigan Math. J.*, 32:237–248, 1985.

[56] J. Marcinkiewicz and A. Zygmund. A theorem of Lusin. *Duke Math. J.*, 4: 473–485, 1938.

[57] K. Minemura. Harmonic functions on real hyperbolic spaces. *Hiroshima Math. J.*, 3:121–151, 1973.

[58] K. Minemura. Eigenfunctions of the Laplacian on a real hyperbolic space. *J. Math. Soc. Japan*, 27(1):82–105, 1975.

[59] Y. Mizuta. On the boundary limits of harmonic functions with gradient in L^p. *Ann. Inst. Fourier, Grenoble*, 34:99–109, 1984.

[60] Y. Mizuta. On the boundary limits of harmonic functions. *Hiroshima Math. J.*, 18:207–217, 1988.

[61] A. Nagel, W. Rudin, and J. H. Shapiro. Tangential boundary behavior of functions in Dirichlet-type spaces. *Ann. Math.*, 116:331–360, 1982.

[62] M. Pavlović. Inequalities for the gradient of eigenfunctions of the invariant Laplacian in the unit ball. *Indag. Mathem., N. S.*, 2:89–98, 1991.

[63] M. Pavlović. On subharmonic behavior and oscillation of functions in balls in $\mathbb{R}^n$. *Publ. Inst. Math. (N.S.)*, 69:18–22, 1994.

[64] M. Pavlović. A Littlewood–Paley theorem for subharmonic functions. *Publ. Inst. Math. (Beograd)*, 68(82):77–82, 2000.

[65] M. Pavlović. A short proof of an inequality of Littlewood and Paley. *Proc. Amer. Math. Soc*, 134:3625–3627, 2006.

[66] M. Pavlović and J. Riihentaus. Classes of quasi-nearly subharmonic functions. *Potential Analysis*, 29:89–104, 2008.

[67] Marco M. Peloso. Möbius invariant spaces on the unit ball. *Michigan Math. J*, 39:509–536, 1992.

[68] I. Privalov. Sur une généralization du théorème de Fatou. *Rec. Math. (Mat. Sbornik)*, 31:232–235, 1923.

[69] T. Ransford. *Potential Theory in the Complex Plane*. London Math. Soc. Student Texts 28, Cambridge University Press, 1995.

[70] J. Riihentaus. On a theorem of Avanissian–Arsove. *Exposition. Math.*, 7:69–72, 1989.

[71] H. L. Royden. *Real Analysis*. Macmillan Publishing Co., New York, NY, third edition, 1988.

[72] W. Rudin. *Function Theory in the Unit Ball of* $\mathbb{C}^n$. Springer-Verlag, New York, NY, 1980.

[73] H. Samii. *Les transformations de Poisson dans le boule hyperbolic*. PhD thesis, Université Nancy 1, 1982.

[74] I. Sokolnikoff. *Tensor Analysis*. Wiley, New York, NY, 1964.

[75] E. M. Stein. *Singular Integrals and Differentiability Properties of Functions*. Princeton University Press, Princeton, NJ, 1970.

[76] E. M. Stein and G. Weiss. *Fourier Analysis on Euclidean Spaces*. Princeton University Press, Princeton, NJ, 1971.

[77] S. Stević. A Littlewood–Paley type inequality. *Bull. Braz. Math. Soc.*, 34:1–7, 2003.

[78] M. Stoll. Hardy-type spaces of harmonic functions on symmetric spaces of noncompact type. *J. Reine Angew. Math.*, 271:63–76, 1974.

[79] M. Stoll. Mean value theorems for harmonic and holomorphic functions on bounded symmetric domains. *J. Reine Angew. Math.*, 290:191–198, 1977.

[80] M. Stoll. Boundary limits of Green potentials in the unit disc. *Arch. Math.*, 44:451–455, 1985.

[81] M. Stoll. Rate of growth of pth means of invariant potentials in the unit ball of $\mathbb{C}^n$. *J. Math. Analysis & Appl.*, 143:480–499, 1989.

[82] M. Stoll. Rate of growth of pth means of invariant potentials in the unit ball of $\mathbb{C}^n$, II. *J. Math. Analysis & Appl.*, 165:374–398, 1992.

[83] M. Stoll. Tangential boundary limits of invariant potentials in the unit ball of $\mathbb{C}^n$. *J. Math. Anal. Appl.*, 177(2):553–571, 1993.

[84] M. Stoll. *Invariant potential theory in the unit ball of* $\mathbf{C}^n$, volume 199 of London Mathematical Society Lecture Note Series. Cambridge University Press, Cambridge, 1994.

[85] M. Stoll. Boundary limits and non-integrability of $\mathcal{M}$-subharmonic functions in the unit ball of $\mathbb{C}^n$ $(n \geq 1)$. *Trans. Amer. Math. Soc.*, 349(9):3773–3785, 1997.

[86] M. Stoll. Weighted tangential boundary limits of subharmonic functions on domains in $\mathbf{R}^n$ $(n \geq 2)$. *Math. Scand.*, 83(2):300–308, 1998.

[87] M. Stoll. On generalizations of the Littlewood–Paley inequalities to domains in $\mathbb{R}^n$ $(n \geq 2)$. Unpublished manuscript, 2004. www.researchgate.net/profile/Manfred_Stoll/publications.

[88] M. Stoll. The Littlewood–Paley inequalities for Hardy–Orlicz spaces of harmonic function on domains in $\mathbb{R}^n$. *Advanced Studies in Pure Mathematics*, 44:363–376, 2006.

[89] M. Stoll. Weighted Dirichlet spaces of harmonic functions on the real hyperbolic ball. *Complex Var. and Elliptic Equ.*, 57(1):63–89, 2012.

[90] M. Stoll. On the Littlewood–Paley inequalities for subharmonic functions on domains in $\mathbb{R}^n$. In *Recent Advances in Harmonic Analysis and Applications*, pages 357–383. Springer–Verlag, New York, NY, 2013.

[91] M. Stoll. Littlewood–Paley theory for subharmonic functions on the unit ball in $\mathbb{R}^n$. *J. Math. Analysis & Appl.*, 420:483–514, 2014.

[92] N. Suzuki. Nonintegrability of harmonic functions in a domain. *Japan J. Math.*, 16:269–278, 1990.

[93] D. Ullrich. *Möbius-invariant potential theory in the unit ball of* $\mathbb{C}^n$. PhD thesis, University of Wisconsin, 1981.

[94] D. Ullrich. Radial limits of $\mathcal{M}$-subharmonic functions. *Trans. Amer. Math. Soc.*, 292:501–518, 1985.

[95] J.-M. G. Wu. L^p densities and boundary behavior of Green potentials. *Indiana Univ. Math. J.*, 28:895–911, 1979.

[96] S. Zhao. On the weighted L^p-integrability of nonnegative $\mathcal{M}$-superharmonic functions. *Proc. Amer. Math. Soc*, 113:677–685, 1992.

[97] K. Zhu. Möbius invariant Hilbert spaces of holomorphic functions in the unit ball of $\mathbb{C}^n$. *Trans. Amer. Math. Soc.*, 323:823–842, 1991.

[98] L. Ziomek. On the boundary behavior in the metric L^p of subharmonic functions. *Studia Math.*, 29:97–105, 1967.

[99] A. Zygmund. *Trigonometric Series.* Cambridge University Press, London, 1968.

Index of Symbols

Index

Printed in the United States
by Baker & Taylor Publisher Services